T0206495

First published in the UK 2019 by Sona Books
an imprint of Danann Publishing Ltd.

WARNING: For private domestic use only, any unauthorised Copying, hiring, lending or public performance of this book is illegal.

Published under licence from Future Publishing Limited a Future PLC group company. All rights reserved. No part of this publication may be reproduced or stored in a retrieval system or transmitted in any form or by any means without the prior written permission of the publisher.

Editor: Hannah Westlake, Designer: Alexander Phoenix

Copy editor for Danann Tom O'Neill

© 2018 Future Publishing PLC

CAT NO: SON0444

ISBN: 978-1-912918-03-4

Made in EU.

JOURNEY TO THE MOON

We've been looking up at the Moon for millennia, but it was always just out of reach. People made up stories about it, associating it with deities or talking of the Man in the Moon who watches over all, while others wondered what it would look like up close.

Then everything changed.

On 20 July 1969, man achieved what had previously been thought impossible: two astronauts landed on the surface of the Moon for the first ever time. More Moon landings followed in a era since dubbed the golden age of space exploration, but our curiosity is far from sated.

In *Journey to the Moon*, find out what really happened when Neil Armstrong and Buzz Aldrin stepped outside their lunar module and uncover what is being done to preserve that moment in history before taking a tour of the craters and maria. Learn how to take the perfect photographs of our lunar neighbour, and discover what's really on the Moon's far side. All you have to do is turn the page to be rocketed into space…

Contents

THE MOON

10

30

50

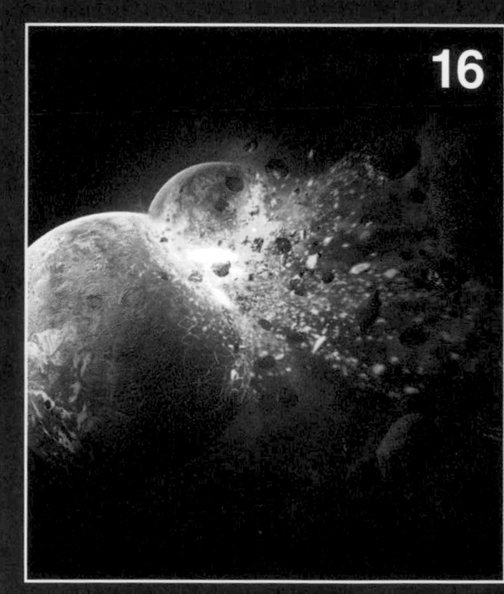
16

24

34

48

132

MOON TOUR

MOON LANDINGS

The Moon

24

16

10

76

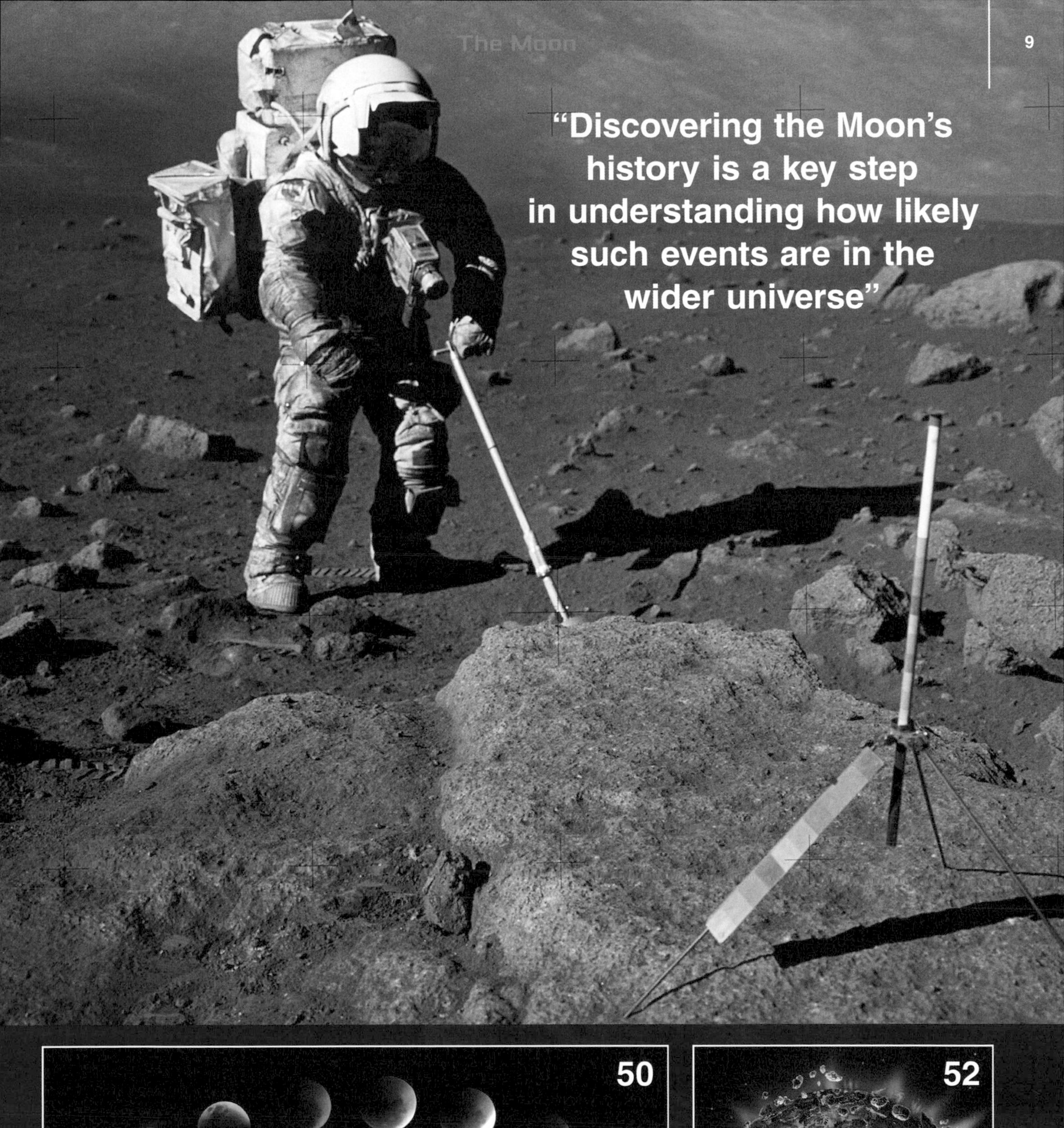

"Discovering the Moon's history is a key step in understanding how likely such events are in the wider universe"

50

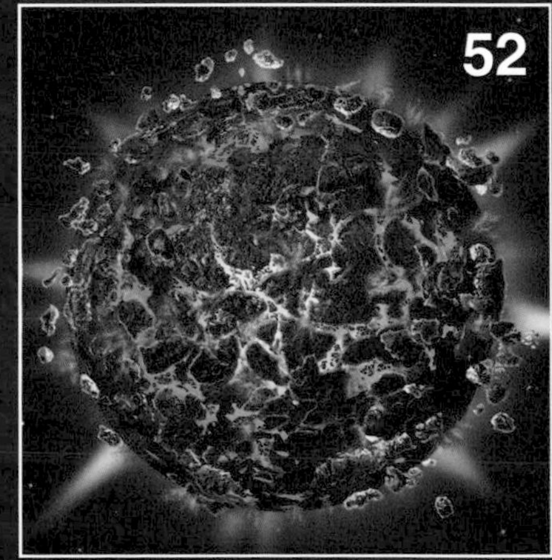
52

Explorer's Guide

The Moon

Take a tour of our nearest neighbour and discover a world much stranger than you thought

The Moon's diameter is 27 per cent of Earth's, which is freakishly large. Jupiter's moon Ganymede is larger in absolute terms, but compared to its parent planet it is just a speck, with less than one-fiftieth of the diameter. The disproportionate size of the Moon reflects its unusual origins. Most other moons are captured asteroids, or leftover material from the ball of dust and gas that coalesced to form the Solar System. But the Moon is made from the rubble of another planet, called Theia, that collided with Earth around 4.5 billion years ago.

When it was first formed, the Moon may have been ten times closer to Earth than it is today and would have caused tides on Earth several kilometres high. Over billions of years, the gravitational drag of the tides has slowed the Moon's orbit, causing it to spiral outwards from us. Earth's gravity also raises small 'land tides' on the Moon, stretching it slightly and creating another drag that has synchronised its rotation, so that it always presents the same face to Earth.

The full Moon in the night sky looks flat, rather than darkening at the edges as you'd expect from a 3D sphere. This effect is caused by the unusual properties of the lunar soil, or 'regolith', which reflects more light when the Sun shines at a low angle, than when it is directly overhead. Moon rock is about the colour of a dry road, but a full Moon appears much brighter than that, because all the tiny shadows cast by the regolith disappear when the Sun shines from behind our viewpoint. The regolith can be up to 20 metres (66 feet) deep on the Moon's highlands and in the early days of lunar exploration there were real concerns that spacecraft would simply sink into it.

Sea of Rain

Schroter's Valley

Copernicus crater

How to get there

1. Lift off
The first 200km (124mi) is the hardest. It takes about 23 tons of fuel to propel each ton of payload into orbit, not counting the mass of the empty rocket stages.

2. Parking orbit
Having accelerated to 7.8km/s (4.8mi/s), the spacecraft makes a couple of orbits around the Earth, a few hundred kilometres up, to check all its systems are working correctly.

3. Transfer burn
To reach the Moon, the spacecraft must fire its engines to accelerate by another 3.2km/s (7,200mph), stretching its circular orbit into a long, thin ellipse.

4. Coast
The transfer burn is timed so that by the time the spacecraft reaches the top of its orbit, the Moon has travelled to the same point in space.

5. Landing
Once it gets close enough, the spacecraft fires its engines again to slow down just enough to be captured by the Moon into an orbit, or even more for a direct landing.

Luna 2 impact site

Sea of Serenity

Sea of Tranquility

How big is the Moon?

The Moon is the fifth largest satellite in the Solar System and the biggest in comparison to its host planet

3,475km (2,160mi) wide

Cleveland

San Francisco

Moon

How far is the Moon?

The Moon is currently 384,400km (238,855mi) away from Earth but is gradually moving away at a rate of 3.82cm (1.5in) a year. If the Earth was a basketball and the Moon a tennis ball — they would be 7.3m (24ft) apart.

Earth

Moon

Top sights to see on the Moon

Seen from Earth, the most obvious features on the Moon are the 'maria' or seas. These are vast plains of solidified lava that originally bled for tens of millions of years from the puncture wounds of asteroid impacts. They are darker than the rest of the Moon's crust because of the iron compounds in the basalt minerals. The Moon long ago cooled and set solid all the way through, so it doesn't have tectonic plates to throw up mountain ranges like Earth's. But asteroid impacts can do the same job in a fraction of the time. Large impacts create mountain ridges around their rims, such as the Montes Rook which form a ring 500 kilometres (310 miles) wide around the Mare Orientale. Where an impact crater hasn't been filled in by lava to form a mare, there is also a central mountain formed where the crust rebounded from the initial impact. Impacts can fling debris out over hundreds of kilometres. These form spokes of lighter asteroid material overlaid on the darker lunar regolith. Tycho crater in the southern hemisphere has rays that extend 1,500 kilometres (930 miles) from its rim.

When the Soviet Union sent Luna 3 to photograph the far side of the Moon for the first time, in 1959, they found it strangely lacking in maria. This is because the crust is thicker on the far side, so asteroid impacts didn't punch all the way through to the magma beneath. Instead the terrain is more irregular, thrust into a jumble of spires by the shockwaves from ancient impacts on the opposite side that travelled through the Moon and burst out of the surface, like an exit wound. Without water or wind, the landscape is unweathered, so every mountain is sharp and jagged.

As well as mountains and craters, the Moon has twisting features that look like river valleys. Called 'rilles', these may have been caused by lava tubes that cooled and sank into the regolith. Hadley Rille, where Apollo 15 landed, runs along the base of the Apennine mountains. Schroter's Valley is the largest lunar rille. It begins at a crater six kilometres (3.7 miles) wide and meanders for hundreds of kilometres in a strip that is ten kilometres (6.2 miles) wide in places. In fact it's so big that it has another smaller rille formed by a second lava flow, running along it like a river meandering through a glacial valley.

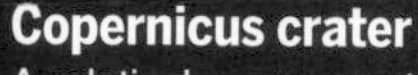

Copernicus crater
A relatively young crater at 'just' 800 million years old. The mountains in the centre are almost as high as Ben Nevis.

Apollo 11 landing site
Chosen because it was fairly flat, Neil Armstrong nevertheless had to manually pilot 6km (3.7mi) from the intended landing site to avoid a boulder field.

Tallest lunar mountain
Mons Huygens is part of the rim of the vast crater that filled with lava to become the Sea of Rains. It is 5.5km (3.4mi) high.

Apollo 17 landing site
The last human footprints on the Moon were left by Eugene Cernan, the commander of Apollo 17, as he climbed back aboard in December 1972.

The lunar orbit explained

The Moon spins on its axis at the same speed as it rotates around the Earth. This 'tidal locking' means a day on the Moon lasts almost four weeks. As the day/night terminator creeps slowly across the surface, we see a different phase illuminated in the sky each night, from new Moon to full Moon and back again.

The Moon's orbit is oval shaped

Perigee
363,400km/225,800mi (distance when closest to Earth)

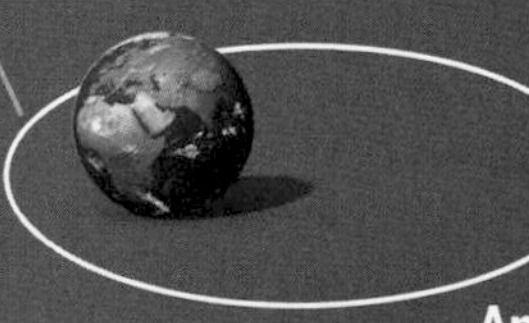

Apogee
405,500km/252,000mi (distance when farthest to Earth)

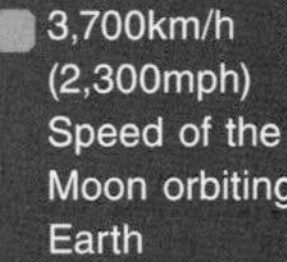

16km/h (10mph)
Speed of the Moon's rotation

3,700km/h (2,300mph)
Speed of the Moon orbiting Earth

27.3 days
The Moon takes about 27 days to complete one orbit

The Moon in numbers

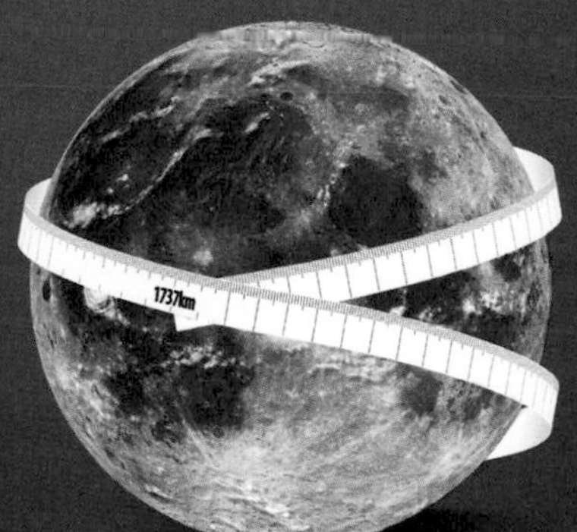

0.1654g
Surface gravity. On the Moon you would be one-sixth your Earth weight

Weather forecast

100°C
-153°C

The Moon has virtually no atmosphere, just a sprinkling of evaporating atoms. So there is no wind to create weather systems. In direct sunlight the ground heats to over 100°C (212°F) but in the shade of a crater it can be as low as -247°C (-413°F).

© Freepik.com; NASA

1,738km
Radius of the Moon — just over a quarter of Earth's

5,000metres
Height of the highest mountains on the Moon — taller than Mont Blanc

Hours in a lunar day — sunrise to sunrise

1972
The last time anyone stepped foot on the Moon

30%
Increase in brightness of the Moon when closest to Earth

135days
to get to the Moon by car travelling at 70mph

The Gravity of the Moon's Situation

Gravity is an ever-present force that shapes the universe, but what about its influences on the Moon?

It has been roughly four and a half billion years since the Moon found the Earth, albeit in a vicious collision. However, the bond between the two is unbreakable and has been unmistakably influential in the evolution of life on Earth. Without gravity, the Moon would have been gone a long time ago, but it is that strange attractive force that has kept the big ball of space rock in our night sky for billions of years.

When the Solar System was just a baby, anchor reined supreme as the planetesimals — minute planets — were gathering material and colliding with the debris that was taking shape. After a few hundred million years, the Moon eventually came into existence and hasn't left since. It is commonly accepted throughout the scientific community that the epoch of the Moon came about when a Mars-sized body collided with Earth, vaporising the planet's crust into space. It was gravity that brought these pieces together, creating a moon that is the largest in the Solar System relative to its host planet. Since then, the Moon hasn't left and Earth's side and has played a rough part in the evolution of the planet and human civilisation.

Although the Moon is about one-quarter of the size and is just over one per cent of the mass of Earth, its gravitational influence can still be felt on our planet — the most evident and obvious example of this is the presence of ocean tides. Ocean water covers about 71 per cent of the Earth's surface and it is all connected. There is a gravitational imbalance between the Sun, Earth and Moon that is the cause of bulges arising on the Earth's surface. The area directly underneath the Moon will experience the attractive force of the Moon's gravity, whereas the opposite will not feel the same attractive force and look to pull material away from the lunar side, causing separate bulges, which are called tidal bulges. Without the tides, life would have never been washed up on the beaches on the young Earth and life as we know it may have never even existed.

In a slightly more insignificant matter, the days on Earth would have been much shorter as well. In a phenomenon known as tidal braking, the friction between the ocean's bulge and Earth's rotation energy is stolen from the Earth's rotation and acts like braking pads in the process. As the Earth rotates faster than the Moon by a much faster rate of 24 hours to 27 days, this drags the ocean bulge in a direction that compensates by robbing the Earth's rotational energy. As a consequence, the length of a day has been slowly getting longer by 2.3 milliseconds per century at the present rate. By this logic, a day-night cycle in the early Carboniferous period roughly 350 million years ago would have been less than 23 hours.

Since intelligent life as we know it has been around over this course of time, exploration of the Moon hasn't been possible until

NASA is actively studying the effects of microgravity, but to a more extreme degree than what is felt on the Moon

half a century ago. It was 20 July 1969 when humans first set foot on the Moon and felt the effects of its gravity up close. Along with Neil Armstrong and Buzz Aldrin, ten other men have experienced the gravity of the Moon over the course of five other Apollo missions.

As the Moon is considerably smaller and less massive, its gravity in turn is much weaker — in fact, it is about 83 per cent weaker than Earth's, meaning that any object on Earth would weigh much less on the Moon. Take one of the astronauts; an average adult male's weight on Earth would be 89 kilograms (196 pounds), but on the Moon, the astronaut would only weigh 15 kilograms (33 pounds)! The common conception of the Moon is that astronauts hop around in the low gravity, looking like if they jumped too hard they wouldn't come back down to the ground! The concept of jumping hard enough in order to break free of a planet's, or in this case Moon's, gravitational shackle is called the scape velocity. On Earth, this velocity is over 11 kilometres per second (equating to roughly 25,000 miles per hour), whereas on our lunar companion the escape velocity is only just over two kilometres per second (4,500 miles per hour). This means less fuel is needed to break free, and hence why the Apollo missions only required a Lunar Module and not a huge rocket.

The receding Moon

If you think a receding hairline is bad, we could lose the Moon!

The aforementioned tidal braking means that we are not just losing time, but also losing the Moon. With tidal breaking, there is a loss of energy from the Earth's rotation and according to the laws of thermodynamics, energy cannot be created or destroyed and just transferred. In this case, the energy is being given to the Moon, which is slowly providing it with the energy to break free of the Earth's gravity.

At the moment, the Moon is receding from the Earth at a rate of four centimetres (1.6 inches) per year. This is roughly the same time in which fingernails grow, and as the Moon is on average 385,000 kilometres (239,000 miles) away from Earth, it is extremely hard to notice on relatively short scales. According to studies and simulations, 4 billion years ago the Moon would have been orbiting Earth at about 130,000 kilometres (80,000 miles) away. If someone was lucky enough to be alive at this point in the Earth's evolution, a full Moon would appear three times larger than it is now.

A modern day full moon would pale in comparison to one 4 billion years ago

The long-term effects of the lunar gravity could prove harmful on the Moon. If ideas of the future do come true, then one day there will be a lunar base. This will be a place on the Moon for humans to build, live and colonise. In preparation for this, scientists have been studying the effects of microgravity on the International Space Station. Throughout the years, they have found that microgravity can have adverse effects including decreasing muscle mass (muscle atrophy) and the slow deterioration of the skeleton. There are also changes to the body's cardiovascular system, eyesight and immune system and these are issues that organisations such as NASA continue to study for human's eventual return to the Moon and beyond.

Armstrong and Aldrin were the first two people to experience the surface gravity of the Moon in 1969

© NASA/Bill Ingalls

What Made our Moon?

Understanding how our lunar companion was formed might just explain how we came to be here

Written by Colin Stuart

© Tobias Roetsch

It is the brightest thing in our night sky. Over the course of history it has been revered as God, trampled on by 12 American men and immortalised in love-soaked poetry. The Moon is our steadfast companion, our only natural satellite as we endlessly orbit the Sun. Yet for an object that has received such scrutiny, arguments still rage about where exactly the Moon came from.

A suitable explanation needs to take into account what is perhaps the Moon's greatest oddity: its size. It is the fifth largest moon in the Solar System, trumping most of the satellites of our much bigger planetary neighbours. In fact, if you compare the size of moons to the size of their host planet, our Moon comes out at the very top of the list. Many of the smaller moons of the Solar System are thought to be captured worlds — bodies that wandered too close to a planet before getting snared in its gravitational pull. Given the size of our Moon, it's hard to imagine that is how it ended up circling the Earth.

As far back as 1879, George Darwin — the astronomer son of the famous naturalist, Charles Darwin — instead proposed that the Earth and Moon were once one body and that the latter formed from material thrown off the spinning Earth. This, he said, would explain why the Moon was moving a little further away from us each year. Supporters of this idea even pointed to the lack of land in the Pacific Ocean — which stretches across half of our planet — as the birthplace of the Moon. However, scientists later realised that any force capable of dislodging such a large amount of Earthly material would likely have destroyed the rest of our planet at the same time.

So attention turned instead to the idea of a giant impact — one that occurred 4.5 billion years ago when the Earth was still forming. It must have been this early because the rocks brought back from the Moon are that old. Astronomers have long believed that the Solar System had a tempestuous infancy, throwing around huge lumps of rock and metal before eventually calming down. What if one of these objects — perhaps one the size of Mars — hit the young Earth, with the Moon forming out of the hot, spinning debris?

On the face of it, this idea makes a lot of sense. We know from the dark patches on the lunar surface that parts of the Moon were once molten. The Moon also has a pretty small iron core — much smaller than the Earth's — and it is less dense than the Earth. This fits, too, because during an impact

"If the Moon was formed from a smashed apart Theia during a blow with the Earth, then it should have its own unique oxygen isotope. Yet it matches the Earth's exactly"

How the Moon wasn't made

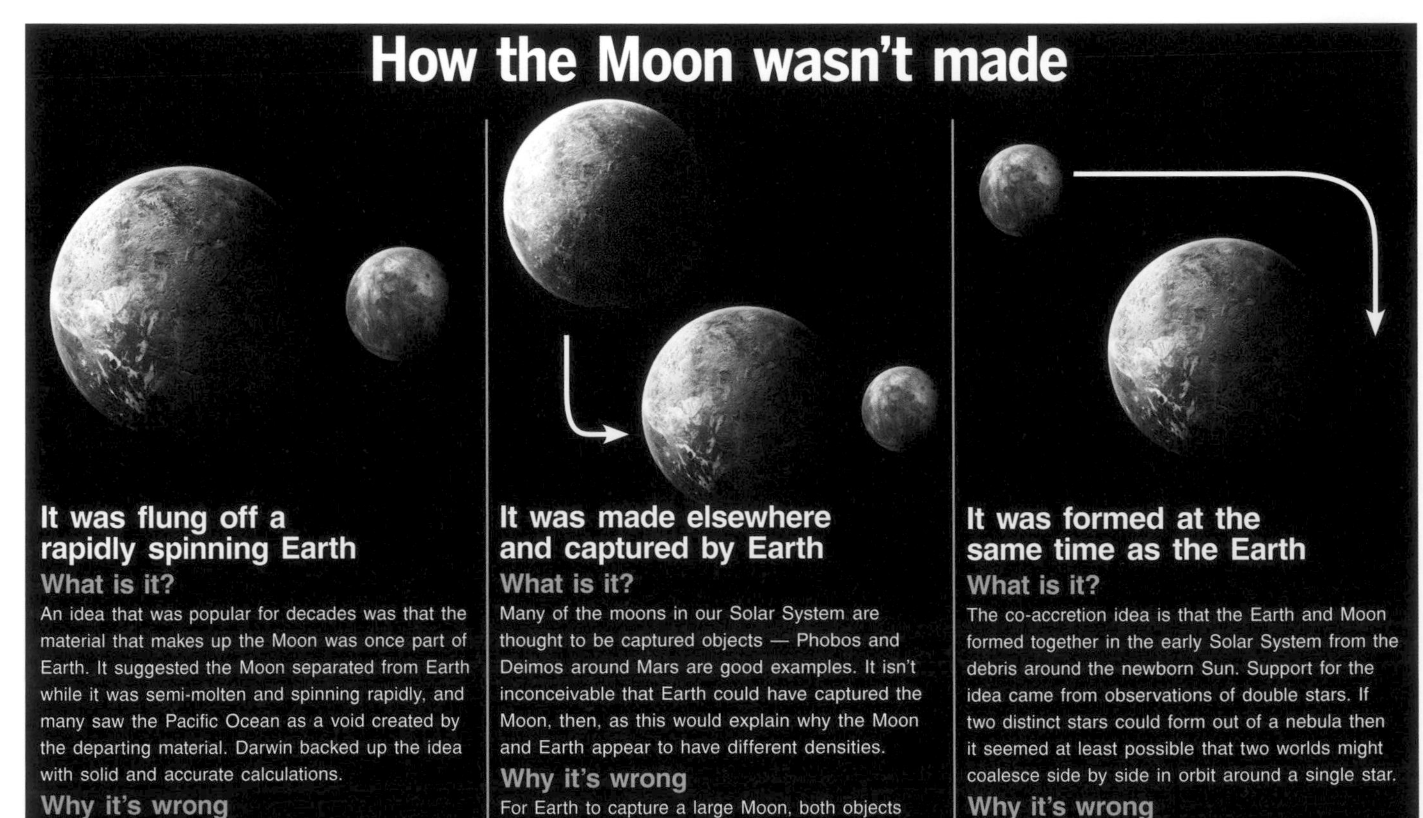

It was flung off a rapidly spinning Earth

What is it?

An idea that was popular for decades was that the material that makes up the Moon was once part of Earth. It suggested the Moon separated from Earth while it was semi-molten and spinning rapidly, and many saw the Pacific Ocean as a void created by the departing material. Darwin backed up the idea with solid and accurate calculations.

Why it's wrong

By the 1930s, calculations showed that Earth would have had to spin at an inconceivable rate to throw off enough material to form the Moon.

It was made elsewhere and captured by Earth

What is it?

Many of the moons in our Solar System are thought to be captured objects — Phobos and Deimos around Mars are good examples. It isn't inconceivable that Earth could have captured the Moon, then, as this would explain why the Moon and Earth appear to have different densities.

Why it's wrong

For Earth to capture a large Moon, both objects would have to travel slowly — a collision was more probable. It is also unlikely that the Earth's gravity would have been able to hold the Moon for so long.

It was formed at the same time as the Earth

What is it?

The co-accretion idea is that the Earth and Moon formed together in the early Solar System from the debris around the newborn Sun. Support for the idea came from observations of double stars. If two distinct stars could form out of a nebula then it seemed at least possible that two worlds might coalesce side by side in orbit around a single star.

Why it's wrong

While the oxygen isotopes may be the same, the densities of the Earth and the Moon and the amounts of iron on each are different.

© Tobias Roetsch

How the Moon was made

1. Theia approaches Earth
A Mars-sized object is on an unalterable collision course with the early Earth.

2. Earth gets hit
The impactor hits the Earth in a head-on collision, vaporising both Theia and the mantle of the Earth.

3. Material is thrown out
The vaporised material from both bodies mixes and is thrown outwards by the huge impact.

4. Debris gathers
Smaller objects begin to condense out of the vapour while continuing to orbit around the Earth.

5. The Moon takes shape
Many of the smaller objects stick together to form a proto-Moon in orbit around the Earth.

6. Our companion is formed
Eventually, all the pieces come together to form the basis of the Moon that we see today.

Moon rock analysis

Hundreds of lunar rock samples were brought back to the Earth during the Apollo missions for scientific study and research

The age of the Moon

Analysis of lunar rocks suggests the Moon is almost as old as the Earth, meaning the collision happened within Earth's first 100 million years.

No sign of water

The Moon rocks show no signs of past interaction with water. All the geology can be explained as rocks being under pressure.

Matching oxygen isotopes

The relative abundances of the three stable isotopes of oxygen are the same on Earth and the Moon, suggesting a common origin.

Differing potassium isotopes

There is slightly more of one particular potassium isotope on the Moon, pointing to it being vaporised during a head-on collision.

Astronaut Harrison Schmitt is seen covered in lunar dirt while collecting samples during the Apollo 17 mission

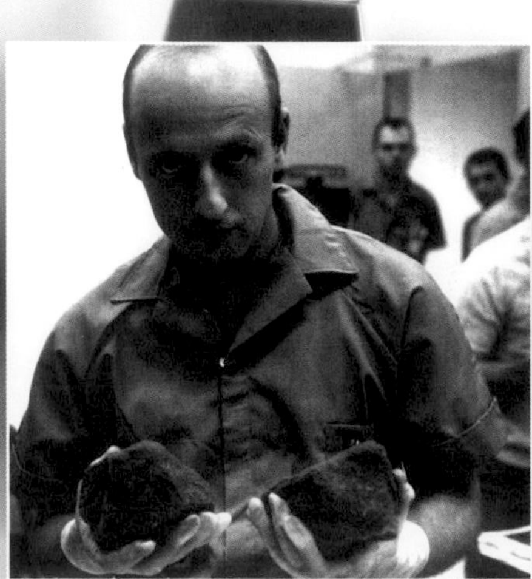

NASA astronaut Charles Conrad, commander of Apollo 12, holds two Moon rocks back on Earth

Wearing special germ-free clothing, Dr Robert Gilruth (right) inspects lunar samples from the Apollo 17 mission

the lightest material would have been thrown the furthest, leaving the heavier stuff here on Earth. Astronomers have a name for this proposed Mars-sized impactor: Theia, named after the Titan who gave birth to the Moon goddess Selene in Greek mythology. And computer modelling has been used to try and figure out what this impact must have been like in order for it to form the modern Moon. Traditionally, the best fit seems to come from a glancing blow — Theia clipping the Earth at an angle of about 45 degrees — and at a relatively slow speed. The debris from the impact, mostly formed from the leftover remnants of Theia, would have formed a ring around the Earth, which then coalesced into the Moon. But recent analysis of Moon rocks returned to the Earth during the Apollo missions appears to throw a spanner in the works.

It is all to do with isotopes. What sets different chemical elements apart is the number of protons present in the nucleus of their atoms. Oxygen, for example, always has eight. Add another proton and you get an entirely different element (fluorine, in this case). But several versions of the same element can exist, each with the same number of protons but a differing number of neutrons. Scientists call these different flavours of the same element 'isotopes'. Oxygen, for example, has three stable isotopes, with eight, nine or ten neutrons.

When it comes to planetary geology, the relative amounts of each of these isotopes present on a celestial object are a key measurement, a bit like a fingerprint. "Each body in the Solar System has a unique oxygen isotope signature," says Dr Kun Wang, assistant professor of geochemistry at Washington University in St Louis. And therein lies the rub. Analysis of the Apollo samples shows that Moon rocks have exactly the same oxygen isotope signature as the Earth. If the Moon was mostly formed from a smashed-apart Theia during a glancing blow with the Earth, then it should have its own unique oxygen isotope signature. Yet instead, it matches the Earth's signature exactly.

Scientists first discovered this as far back as 2001, but many researchers believed that this apparent similarity was just an artefact of the precision of the experiments — that one day more accurate analysis would be able to show that there was actually a tiny difference after all. But the latest research has found that even with much more precise measurements, the oxygen isotope signature is still identical, and therefore we know that the Moon cannot have come from Theia alone.

Wang believes this points to a much more violent collision, one that melted the outer layers of both Earth and Theia. This material then mixed together to form a vapour — a cloud of material — stretching from our planet

Mars-sized Theia approaches the still-molten Earth before the head-on collision

Theia by numbers

6,000km

The width of the Theia impactor, which is about the same size as Mars

60-80°

The axial tilt of the Earth after Theia collided with the planet

2000

The year that the name Theia was proposed by English geochemist Alex Halliday

1974

The year the giant impact hypothesis was first presented at a conference

45°

Although new research suggests a head-on collision, the old picture had a 45-degree glancing blow

4.31 billion

The number of years ago it is thought Theia collided with the Earth to form the Moon

out to 500 Earth radii. The Moon then condensed from this cloud, explaining why both bodies now have the same oxygen isotopes. "Once they mix together, it doesn't matter what the oxygen isotopes of the two bodies were before," says Wang. But if the notion of a more catastrophic impact is to be accepted, it needs more than one strand of supporting evidence. And so that is exactly what Wang set out to find.

He analysed seven different Moon rock samples from multiple Apollo missions, along with samples of Earth rocks, measuring the different abundances of isotopes of potassium using a technique ten times more accurate than previously possible. Along with his colleague Stein Jacobsen from Harvard University, he published his results. He found that the Moon rocks had a greater abundance of one particular potassium isotope at the level of 0.4 parts per 1,000 more than the Earth. "Potassium is a lot more volatile than oxygen, meaning it is more likely to vaporise and be mobile after the collision," says Helen Williams, an Earth scientist at the University of Cambridge, UK. So the potassium was more likely to end up far away from the Earth and become incorporated as part of the Moon. But for potassium to be vaporised in the first place, the collision must have vaporised both Theia and much of the Earth's surface. To Wang, that has all the hallmarks of a head-on collision rather than a glancing blow.

But even if he is right, there are still some outstanding Moon mysteries in need of explanation — none more so than the unusual tilt of the Moon's orbit around the Earth. The Moon would have initially formed in an orbit matching the orientation of Earth's equator, and then as it moved further from our planet, the gravitational pull of the Sun would have forced it into line with the orbits of the other planets — a plane known as the 'ecliptic'.

Yet today's Moon orbits at an angle of five degrees to the ecliptic. "That might not sound like much, but all the other big moons of the Solar System are inclined at less than a degree to their planets — so the Moon really stands out," says Douglas Hamilton, professor of astronomy at the University of Maryland. A team led by Hamilton has recently attempted to explain this strange anomaly. They ran many computer simulations of the giant impact, with slightly different parameters each time. The one that gave the closest match to the Moon's current orbit suggests that Theia's impact was actually a lot more calamitous for our planet than previous models have suggested.

The almighty wallop from Theia would have sent the Earth spinning much faster — more than twice as fast, in fact, as other previous models have suggested. What's more, the Earth would have been knocked over almost on its side, with its axis tilted somewhere between 60 and 80 degrees to the ecliptic (today it is only tilted by 23.4 degrees).

This high inclination affected the Moon as it retreated from the Earth, forcing it into an orbit tilted at an angle of around 30 degrees to the ecliptic. "It then settled down to five degrees over the last 4.5 billion years," says Hamilton. At the same time, the Earth's axis started to straighten up to its present position. It just goes to show that our ideas about the formation of the Moon are still very much in flux. Quite how we came to have such a large Moon on an inclined orbit is a puzzle still occupying teams of astronomers around the world. But it seems we are getting closer.

And that's important, because discovering the Moon's history is a key step in understanding how likely such events are in the wider universe. And, in turn, that might help us answer a much bigger question: whether we are alone in the universe. That's because many scientists have speculated that the churning of the oceans by a Moon that was much closer to the Earth than it is today could have played a key role in the early development of life on Earth. Its gravitational pull also stabilises the Earth's axis, keeping our seasons steady and reliable.

This flurry of recent research has put us one step closer to understanding how our Moon came to be, and it may one day help us understand our place in the universe.

We are still not certain how the Moon ended up in orbit around the Earth

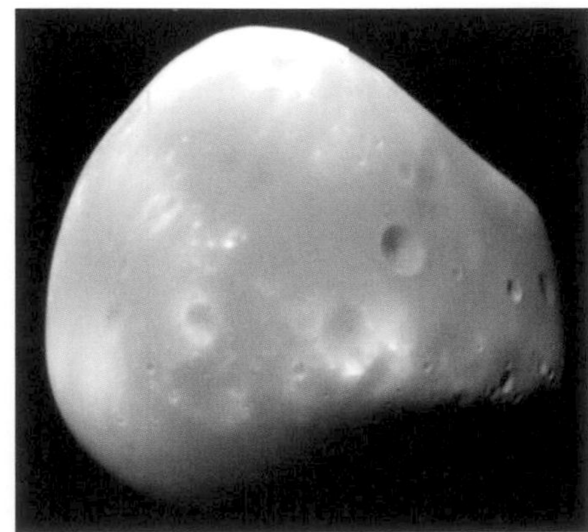

Unlike Mars' moon Deimos, our Moon wasn't captured as it passed by our planet

Moonset as seen by astronaut Tim Peake aboard the International Space Station

Lunar make-up

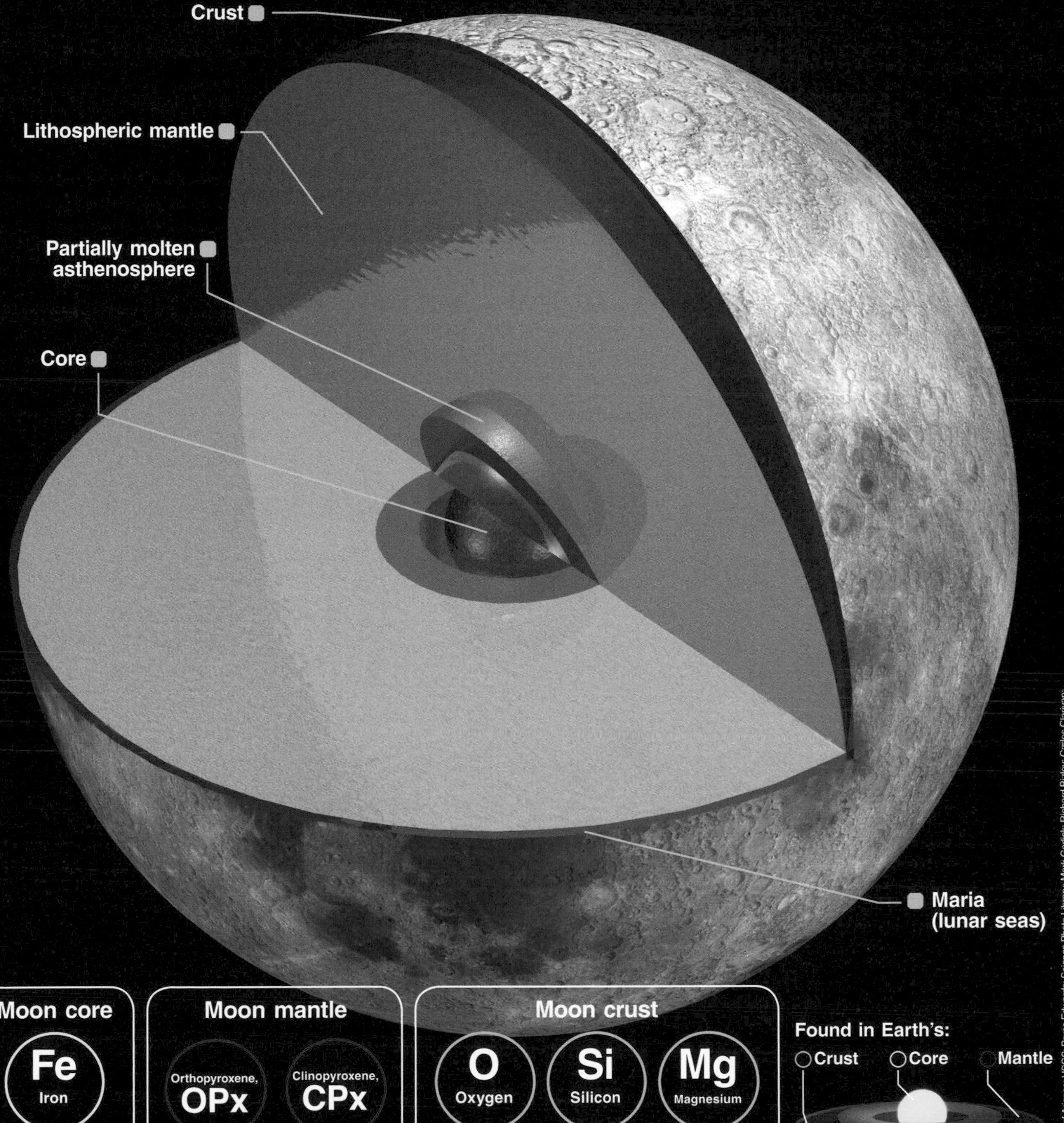

Moon core

Fe Iron

Ni Nickel

Moon mantle

Orthopyroxene, OPx

Clinopyroxene, CPx

Olivine

Moon crust

O Oxygen

Si Silicon

Mg Magnesium

Fe Iron

Ca Calcium

Al Aluminium

Found in Earth's:

Crust

Core

Mantle

© NASA; JPL-Caltech; University of Arizona; USGS; Bryce Edwards; Science Photo Library; Mark Garlick; Richard Bizley; Carlos Clarivan

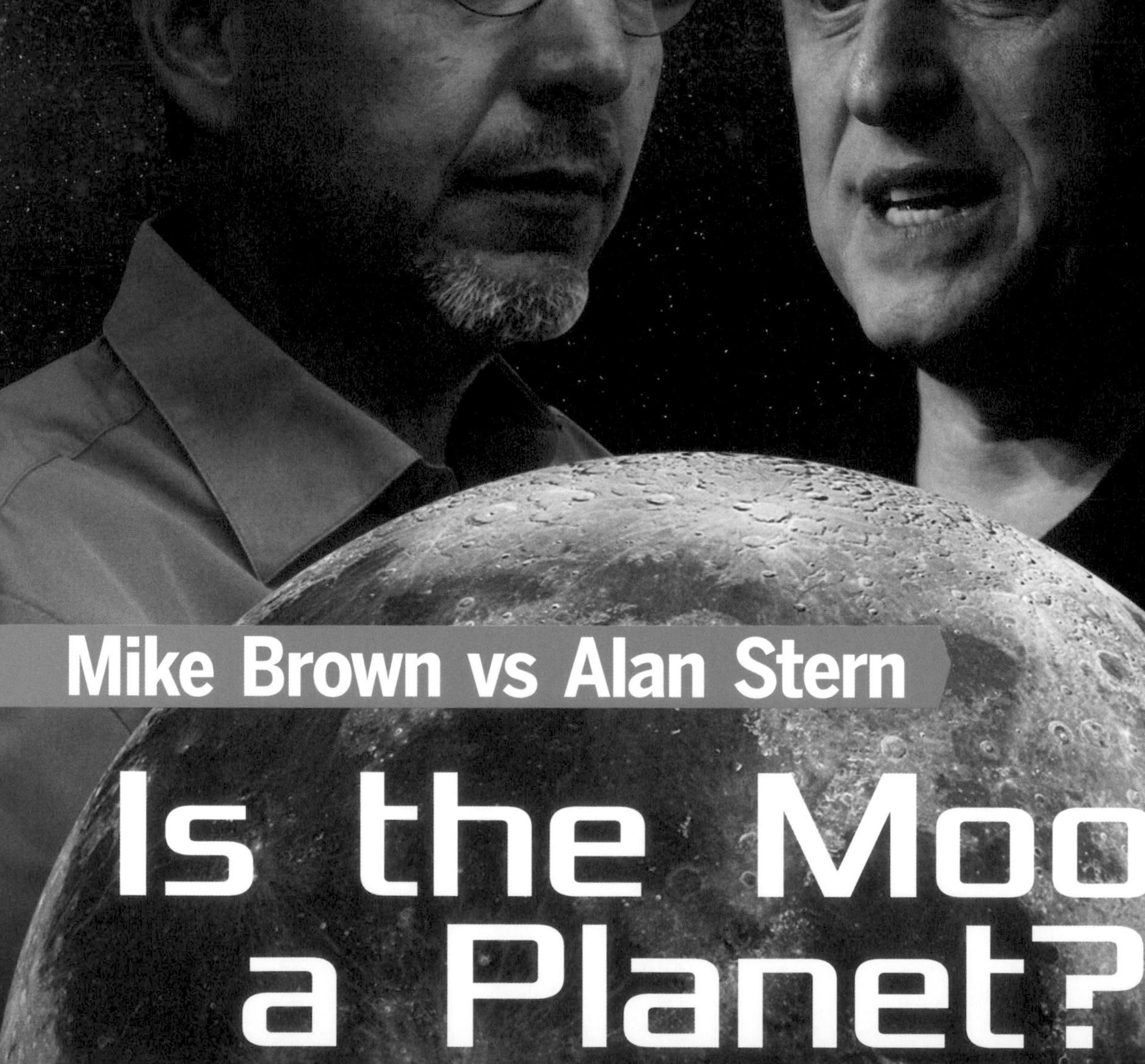

Mike Brown vs Alan Stern

Is the Moon a Planet?

New Horizons' scientists have fuelled debate by creating their own definition of a planet, but it also has consequences for our Moon

Interviewed by David Crookes

Scientists on the New Horizons mission to Pluto have long expressed their frustration at the body's demotion to a dwarf planet. On 24 August 2006, the International Astronomical Union (IAU) decided to nail down the definition of a planet but it controversially demoted Pluto, sparking an argument that continues to rage to this day.

According to the IAU, a planet needs to be round, orbit the Sun and, crucially, clear the neighbourhood around its orbit. Yet Pluto shares its orbital neighbourhood with Kuiper Belt objects and it crosses Neptune's path, placing it into what was then a newly created category of "dwarf planet" along with Ceres and Eris. Alan Stern, who has led New Horizons, has been vocal about Pluto's "miscarriage of justice" for many years and now he and a group of scientists have devised their own definition. They presented it at the Lunar and Planetary Science Conference but it came with consequences: it also defines moons as planets. This now means there are two definitions, but which is right? Can moons be planets? Caltech's Mike Brown doesn't think so, believing the IAU definition to be correct. He calls himself the 'Pluto Killer' and is aghast that "the stupid Pluto stories are back", as he wrote on Twitter. But Brown, who co-discovered Eris in 2005, predicts there is a large planet in the outer Solar System 5,000 times the mass of Pluto.

That, he says, is Planet Nine, claiming Pluto will always be a dwarf. Here, Stern and Brown tell All About Space just why they hold their differing opinions.

It's getting on for 11 years since the IAU downgraded Pluto's planetary status. But does the definition of a planet still make perfect sense in terms of the science that we know today?

Mike Brown (MB): Yes. I would say we have learned absolutely nothing new about what a planet is and nothing that would lead you to have to redo a definition. It's not impossible: we could make new discoveries that would challenge our current concept, but none of those have been made so far. My interpretation is that those who want to redefine it feel like Pluto is in the news a lot these days and that this is their last chance.

Alan Stern (AS): The IAU definition of 2006 is not only antiquated but it was developed in a rush by a bunch of scientists from another field: astronomy. But astronomers and planetary scientists are as different in terms of expertise as, say, neurology and podiatry in medicine. I know that as a planetary scientist, I have very little expertise in black holes and galaxies. Similarly, astronomers have very little expertise in real-world planets. But back in 2006, they really botched it up and they have created some headaches for educators, school children and the public who say, "what the heck, it doesn't add up. Every sci-fi planet I've ever seen looks like Pluto. How can it not be a planet?" The new definition actually works much better with the things that we know and one of the nice attributes of it is that it is actually designed by working experts in planetary science.

But isn't the IAU's definition now set as the one that will be referred to by most people?

MB: Yes. Alan will argue until he can't talk any more that Pluto should be a planet and then people will see his argument and say, "oh, is it going to be a planet again?" and the answer is no. There is no groundswell of movement to make it a planet again, just a very vocal minority of people who want it to be so who will continue to be vocal.

AS: It's still open. Why am I getting interviewed five or ten times a week on this topic, years after the IAU vote? In the scientific community, why are papers being written on it? Mike is trying to say, "don't pay attention, it's a settled matter", because he hopes to keep the status quo. You'll find planetary scientists who agree with Mike but I think you'll find a great majority do not. You can search the literature on Google Scholar, look up Pluto in technical papers and see the word planet being used by my colleagues routinely. And you can do the same for these satellites, the moons of the planets and other worlds in the Kuiper Belt. It's just data. You don't have to ask for opinion. Just find what's published.

Much of the definition and downgrading of Pluto appears to hinge very much on whether a body is able to clear the neighbourhood around its orbit. But how crucial is that?

AS: It's not important whatsoever: that's only about where an object is and what is next to it. So, you know, in geology — to make an analogy — we don't classify mountains according to whether or not they are isolated or come as a group, or whether they are in a linear range or any other

"In biology, we don't decide whether a cow is a cow based on whether it is in a herd or isolated" Alan Stern

After months of testing and a 9.5-year journey over 4.8bn km (3bn mi), NASA's New Horizons craft made its closest approach to Pluto in July 2015

association with regards to what is next to them. Similarly, in biology, we don't decide whether a cow is a cow based on whether it is in a herd or isolated. An object is really about its own attributes and not what it is next to. We don't classify stars according to whether they are in groups or galaxies or not, or whether they clear any orbit in a galaxy (and in fact none of them do). We don't classify asteroids and galaxies that way either. But the astronomers of the IAU do this, in their very flawed planet definition, to limit the number of planets to a manageable level, which is quite unscientific.

Data is data. If there are a lot of mountains, so be it. If there are a lot of rivers or species or hundreds of scientific elements, so be it. We don't try to manage the number to be small. Astronomers don't try to manage the numbers of galaxies or stars or any other type of objects in the universe except one: planets. And this disastrous definition, which no one is happy with years later, means they still have to pay for it in terms of their reputation. The controversy will not go away because they botched it so bad.

MB: You know, I would say the definition that the IAU adopted is poorly worded so I won't defend that, but I will defend the concept that they were trying to describe. It really is very simple: if you look at our Solar System with fresh eyes, it is very difficult not to say, "oh wow, look, there are eight things that are large and they gravitationally dominate everything else that gets around them." So you call that clearing the Solar System, or you call it something else. But if you miss that simple most profound fact about the bodies in the Solar System, then you've kind of missed what the Solar System is all about. That is what the IAU is trying to describe and that is why it does matter. There is such a difference in our Solar System between these eight bodies and how they got there and why, and all the other tiny bodies are flitting in and out or going around these bodies.

"This is just a nostalgic, desperate attempt to get Pluto to be a planet again and moons are sort of the collateral damage" Mike Brown

Mike Brown is known as the 'Pluto Killer' for his involvement in Pluto's demotion from the status of planet to dwarf planet

In reports about defining what is a planet, Alan has said that Pluto should be upgraded along with the Earth's Moon, two moons orbiting Jupiter and two circling Saturn. The argument is that a planet should be defined by a body's intrinsic physical properties rather than their extrinsic orbital properties. Are you able to elaborate?

AS: Very simply, we recognise whether or not something is or is not a planet based entirely upon its own characteristics and not what it is near to. So, for example, large moons of planets are recognised that way and, what's more, we have recognised satellites of planets that are themselves planets, historically, for centuries. If you do some Google searches, you'll see that professional planetary scientists call Titan and Europa by the name planet. You'll see it if you go to scientific meetings. It's the way that we describe these things and you'll find these references throughout the 21st and 20th centuries.

Is there great merit in this definition?

MB: No. This is precisely the argument that we had 11 years ago and it was rejected. It will get attention because it is Pluto and people love the idea of people fighting about Pluto, but it is not a good idea because it ignores the Solar System. Some people say classification doesn't matter but I would argue the opposite as the way you classify things is what drives the questions that you ask.

And so the question that we ask in the Solar System is: how did the planets form? When we ask that question, we're not asking about moons or tiny bodies — we are separating out these planets from all other small bodies. We then ask: why there are planets and small bodies? Why are there moons? Nobody asks the question: why are there round things? And that's because we know the answer to that. That's just gravity.

I think, finally, Alan has admitted that this [definition] has to include the Moon. For many years, they tried to have it both ways: they wanted to say, "Everything round is a planet, except for moons." And I would say, "You just said that it doesn't matter what it is, so how come the Moon is not a planet?" Now they have to admit that this definition makes the Moon a planet, and then that just makes it silly. There is nobody on Earth who is sad because the Moon was declared to not be a planet 500 years ago. We've moved on and it seems crazy to go back to it.

AS: It's actually a very symmetrical definition to the way that we treat stars, asteroids, galaxies and other objects in space. We have satellite galaxies that are galaxies; we have satellite asteroids that are called asteroids; and we have binary stars that are both stars — one goes around the other and even though one is smaller than the other, we call them both stars. And so these big round worlds with surface areas that are large are routinely called planets, and one of the things I like best about this definition is that it is well aligned with other classification schemes. So asteroids can orbit asteroids, stars can orbit stars and, lo and behold, planets can orbit planets.

What about the argument that has been put forward that says bodies orbiting other planets and not just the Sun could be a method of determining whether moons could be upgraded?

MB: It would include the Moon. It would include four moons of

Would New Horizons still have got off the ground if Pluto were defined as a dwarf planet during funding?

A composite of enhanced colour images of Pluto (right) and Charon (above), taken by New Horizons in 2015. The lead of the mission, Alan Stern, has created a new definition of planets

> "That definitely actually is very similar to the geophysical planet definition that we are putting forward" Alan Stern

Jupiter and it would include at least Titan, probably more, and actually a lot of the moons of Saturn. So there would be a dozen or more moons that would suddenly be called planets. Another thing that just strikes me as semi-ridiculous about this proposal is that if this were important, if moons should be planets, how come nobody proposed this until Pluto was demoted?

This is really not about moons being planets, which is just sort of an aside that has to happen too. This is just a nostalgic, desperate attempt to get Pluto to be a planet again and moons are sort of the collateral damage of the desperate attempt.

One of the arguments against the 2006 definition is that it only recognises objects orbiting our Sun...

AS: Yes, that definition excluded planets around other stars and objects orbiting freely in space. Our definition takes that all under the wing if you will and handles all of those cases very simply.

MB: That part of the definition is often misstated and I think purposely misstated. The IAU definition says that we are going to define planets in our own Solar System and that we are declining to yet make a definition for things outside of our Solar System because we don't know enough at this point. That's a reasonable thing to do but people, I think, have purposely misread that to say, "oh, they say there are only planets around our Sun and not around other stars", and that is an attempt to confuse people.

The argument against the IAU definition also says no planet in the Solar System can clear an orbit because small cosmic bodies fly through them. Is that a valid point?

MB: So again, these are all the arguments that were put forward to try to confuse people. Obviously, the astronomers who were voting on the definition of planets knew what they were talking about and so what they meant when they said clearing the orbit clearly meant clearing of all of the other major bodies — there are always going to be smaller bodies there. I agree that the definition is poorly stated but the concept is rock solid. People have just been trying to take the definition apart and say we need to classify Pluto as a planet again. The definition could certainly be stated better but it's still right.

AS: It's true, you know, near-Earth asteroids surround the Earth. Jupiter has the Trojan asteroids and Pluto crosses Neptune's orbit, and part of the flaw in the IAU definition is that if you take it literally, which is what we do in science because we have to be precise, then it rules out all of the planets in the Solar System because there is not one that doesn't have other objects around it. So they did a poor job and we're trying to clean that up.

Before the 2006 decision, there had been a proposal to include 12 planets and that would have meant Pluto's status remaining as it was, together with the addition of Ceres, Eris and Charon. Were there some valid points in that argument?

MB: It was a weird convoluted attempt at a definition to keep Pluto a planet. I was actually unhappy with that definition because they were trying to make it seem like it was not a big change. They said round things are planets but the only ones that count are Pluto, Charon and Eris, which ignored the other 200 round things that we know about in the Solar System because people would have found that to be a little shocking. So they had a definition but then they didn't believe their own definition enough to

How the International Astronomical Union defines a planet

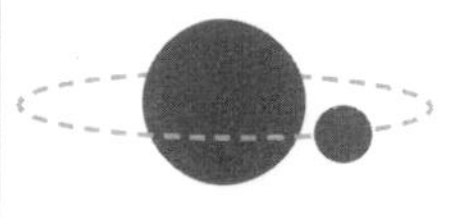

It must orbit the Sun
The Sun is the centre of the Solar System and it pulls planets into a curved orbit.

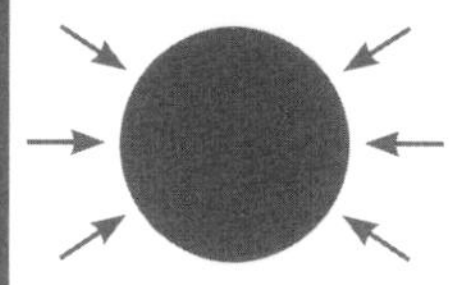

It has to be round (ish)
A body needs sufficient mass so that its own gravity squashes it into a nearly round ball.

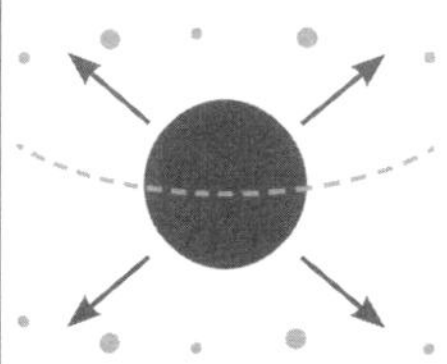

It needs its own space
It must "clear the neighbourhood around its orbit" — that is, it must be gravitationally dominant with no other comparably sized bodies in its vicinity.

What the alternative definition states

Alan Stern and other planetary scientists have put forward their own definition of a planet. It states:

> *Planet: a sub-stellar mass body that has never undergone nuclear fusion and that has sufficient self-gravitation to assume a spheroidal shape adequately described by a triaxial ellipsoid regardless of its orbital parameters*

In other words, it cares less about the orbit of a body or the gravitational effect it may have on other objects and concentrates entirely on what it intrinsically is. It means Pluto along with Titan, Charon and our Moon are defined as planets.

talk about what was really going on there. It's entirely possible that whole decision-making process could have gone differently if they hadn't made so many mistakes in trying to roll out how it went. And in the end, I think people just got irritated with the desperate attempts to keep Pluto and just said forget it. Let's just finally have the correct scientific definition and forget all of the nostalgia that we need to keep Pluto around.

AS: Very simply the IAU put together a committee of experts in their field and they worked on it pretty hard. They weren't bending over backwards or doing anything else and that definition actually is very similar to the geophysical planet definition that we are putting forward.

When Pluto was demoted, it meant that other bodies were too. Mike, you co-discovered Eris and had a lot to lose at that time. Did you feel any emotion when it was classified as a dwarf planet?

MB: I have to say, I was shocked and pleased when the decision was made because I knew it was inherently right. I was watching the decision on some live stream on TV and when the vote came in, I was elated. It was a hard decision for astronomers to make but it was absolutely the right one. I called up my wife and said, "They just did it; they actually made the right decision. Pluto is not a planet anymore." And she said, "Does that mean Eris is not a planet?" And I said, "Er, yes." There was a little sad part of me for Eris but it was completely the right thing.

Rethinking what is or isn't a planet isn't a new thing. In 1801, Ceres was thought to be the eighth planet and it remained that way for half a century until it was reclassified as an asteroid, and then it was upgraded to the status of dwarf planet.

AS: Exactly. In science, we learn more and ideas evolve. We are open to new data and ideas and boundaries move back and forth. In the 20th century, we only knew of nine planets and you could memorise their names. But things have changed a lot since the 1990s when many planets started being discovered around other stars. Since then, we've found distant worlds beyond Neptune and we know every star that we look at has planets. So it's an old, antiquated, 20th-century view that we should be able to name all the planets. Much like we do with mountains and rivers on Earth and the stars in the sky, we just catalogue them. I think it's wonderful that we are discovering more of them and the public get it. It's like Star Trek, there are a crazy number of them but you have to give up naming them all.

Have these changing definitions affected science in any way — perhaps by making certain bodies less attractive so you're not able to get as much funding for a study, for instance?

AS: No, I don't think so. NASA's most recent selection of missions — $1 billion worth of space exploration — includes Lucy and Psyche on missions to large asteroids: items that are clearly not planets. They are fully funded because of the importance of them.

MB: You have to justify missions based on good scientific arguments, not by trying to pretend something is something it's not. If Eris, for example, had been declared one of ten planets, it would get an inordinate amount attention and funding and inordinate other things and that would be crazy. We need to explain why they are important to study and that they don't have to be planets to be interesting.

Do you think New Horizons would have got off the ground if Pluto had been classified as a dwarf planet when the mission was proposed?

AS: I rather suspect it would have been funded. Nothing about Pluto would change simply by changing the nomenclature. Nonetheless, the nomenclature is today antiquated and wrong and I really appreciate it when journalists tell that story in a fair way. I think the tide has turned and even in textbooks now it is clear they are backtracking from the flawed IAU definition.

MB: It's not impossible. A lot of the justification they had early on was that this was the mission to the last planet; but I think that we now know enough to still make the argument that we'd like to go visit Pluto. But it would have been harder and I think that's okay. We need to work hard if we are spending $1 billion of taxpayers' money to go fly a spacecraft out there.

Does a definition that opens the way for many more objects to be defined as a planet mean that new discoveries are devalued in some way?

MB: 11 years ago, I would not have guessed there was a possibility that there was another planet out there in the Solar System. In fact, when the new IAU definition was made I think I was clear that this closed the door on Solar System planets. Now I think I was wrong. There is evidence of a Planet Nine that is 5,000 times more massive than Pluto and when it is found, I think it will get the appropriate attention as an actual, important major body. But it also shows how important it was ten years ago to solidify our definition

Could the Moon be defined as a planet? The alternative definition suggests so, while the IAU's definition does not

© Shutterstock; Getty Images; Win McNamee; Don Bartlett; NASA; ESO; Claus Madsen; JHUAPL; SwRI; NOAA

"When the IAU definition was made, I was clear that this closed the door on Solar System planets… I was wrong" Mike Brown

of the word planet. If we had 200 planets out there and we said, "We've found another one", people would say, "That's no big deal, there are 200."

AS: I don't look at it in terms of valuation. We don't devalue stars just because we're finding new ones. We don't devalue the new discovery of species and there are thousands of species on Earth. When a new element is discovered, we don't say it is 115 times less important as hydrogen, the first element. This is an anti-scientific approach that says somehow, because you find more planets, they become less and less important. There is no analogy to it anywhere else in science. We are scientists and science needs to be informed by data.

Observer's Guide to the

Supermoon

Make the most of the lunar surface as the Moon makes its close approach to the Earth

© Stephen Spraggon; Alamy

Enthusiasts of observing our nearest celestial companion always look forward to the rise of what's called the supermoon. This phenomenon occurs when our lunar companion makes its closest approach to Earth on its elliptical orbit, which is known as perigee, when it's also in its full phase.

At such a time, the Moon can look larger and brighter than it normally does — especially when it is seen rising above the horizon. This is an optical illusion, which causes our lunar companion to seem bigger than it really is when it is high in the sky.

With the effects brought about by our celestial sphere aside, the supermoon will appear seven per cent bigger than a standard full Moon and around 12 per cent larger than when the Moon is at apogee — the point in its elliptical orbit when it's furthest from the Earth.

Whether you're observing the surface of the Moon with a telescope or binoculars, or simply enjoying the spectacle with the naked eye, a supermoon makes an enjoyable sight to behold, with some astronomers reporting that the slight increase in size allows them to observe some of the lunar surface's finer details with ease. On the evenings where this astronomical phenomenon takes place you'll find that the Moon will appear brighter by quite a few per cent, washing out craters, lunar maria and other fascinating features. To remedy this, we recommend that you use a Moon filter with your telescope so that you can knock down brightness and boost contrast through the field of view.

While a full Moon is perhaps the worst time to gaze upon the lunar surface, there is a plus side: it does give you the most comprehensive view of the Moon's litany of maria — sprawling, dark regions more commonly known as seas. These vast plains do not contain any water, but were once oceans of lava present in our satellite's younger days. Along with the Sea of Tranquility — the site of the historic first Moon landing — there are seas of Cleverness, Nectar, Clouds and many others. Smaller plains are known as 'lacus' or lakes, with wistful names like the Lakes of Softness, Dreams, Perseverance and Solitude. Other related features include the 'sinus' or bays (such as the Bays of Rainbows, Roughness and Dew) and 'palus' or marshes (including the Marshes of Decay and Sleep). The largest of these features are visible with the unaided eye and can be enhanced with a pair of low-power binoculars.

After the lunar seas, it's the craters you'll notice next — huge pits created when space debris piled headlong into the lunar regolith. These craters come in two basic types: simple and complex. Complex craters boast an additional central peak, and one of the most popular and accessible of these is Tycho, which is also one of the youngest. Just over 100 million years ago, the southern area of the Moon was struck and the

How to spot the Apollo 11 Moon landing site

01 Find the centre of the Moon

Begin right in the middle of the Moon and move up the centre line until you're about level with the crater Copernicus.

02 Move to the right

Move your gaze over to the right until you come across a big dark sea. It should have another sea of about the same size joining it to the top left.

03 Find the Sea of Tranquility

The bottom of the Sea of Tranquility is split into two sections. The Apollo 11 landing site is the left-hand area.

04 Locate the approximate landing site

Although you won't see any detail due to the landing site itself being very small, it is located around 20km (12.4mi) south-southwest of the crater Sabine D.

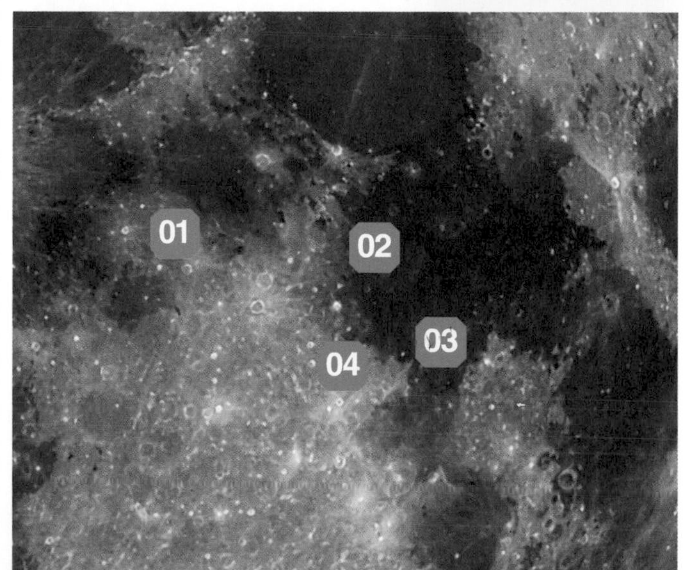

What causes a supermoon?

When a full Moon coincides with perigee — the point in its orbit where it is closest to Earth — our Moon appears slightly larger than usual. Here's what happened in 2016

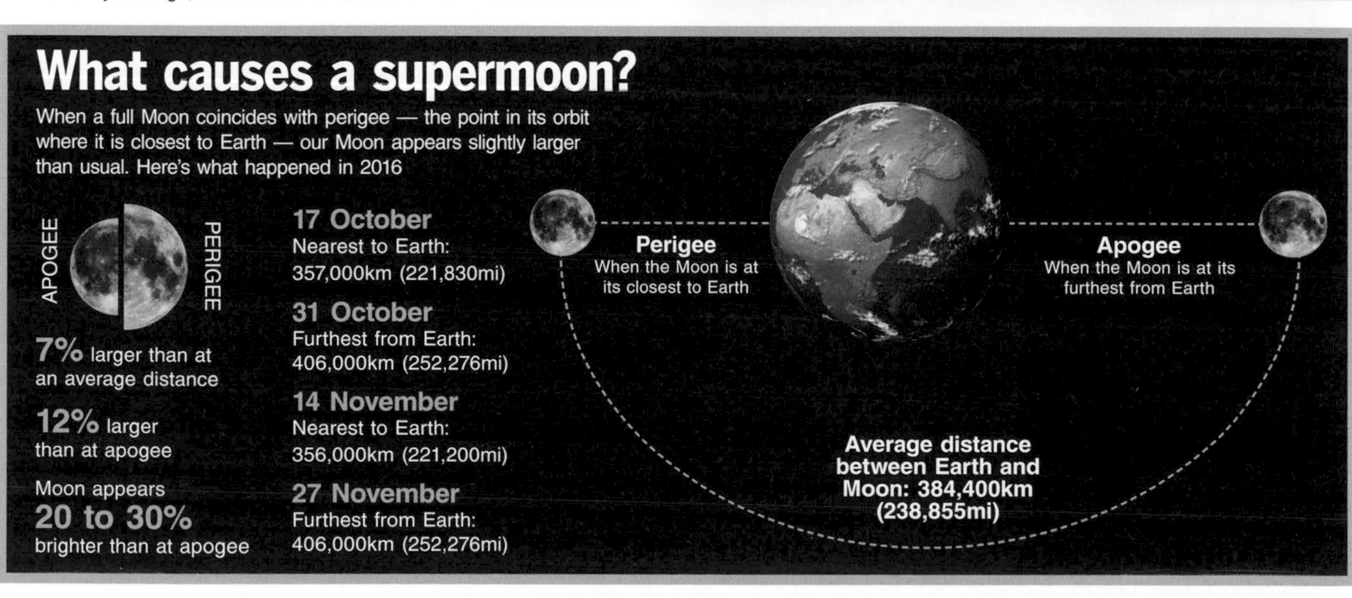

energy of the impact melted some of the rock, throwing it high into the lunar sky. Instantly hitting ice-cold space, the ejecta solidified into glass beads, which then fell back to the surface. If you look closely in the area around Tycho, you'll see long, thin 'rays' stretching outwards like the spokes of a wheel and those glass beads glinting in the sunlight.

Slightly further towards the lunar limb you'll find the crater Clavius. Consisting of one large, old crater whose floor is peppered with holes from later impacts, it shows that the Moon was hit at many points during its history. Depending on the time of the month you are looking, you might also see shadows stretching out like tentacles on the crater floor; they are being cast by the towering rim of the crater, which forms a Moon mountain range that is several kilometres high. Even higher mountains are found around the edge of maria, with the largest being Mons Huygens (at 5.5 kilometres or 3.4 miles tall, it is more than half the size of Mount Everest here on Earth). The famous Italian astronomer Galileo Galilei was able to use these shadows to work out the mountain heights for the first time.

Other popular targets for amateur Moon-watchers are volcanic rilles. While their exact origin is unclear, they are likely either ancient transport routes for the Moon's bygone lava flows or cracks in the lunar crust. One of the most famous rilles is the 100-kilometre (62-mile) long Rupes Recta (also known as the 'Straight Wall'), which forms part of the Mare Nubium not far from Tycho in the Moon's southern hemisphere. Moving to the northern hemisphere, you'll also find Hadley Rille near Mons Hadley in the rugged Montes Apenninus mountain range. It was here that the Apollo 15 astronauts placed a small aluminium sculpture known as 'The Fallen Astronaut', in honour of those who had lost their lives in space exploration endeavours.

However you choose to view the supermoon, whether you go hunting for seas, lakes, bays, craters, marshes, rays, mountain ridges or rilles, or if you prefer to watch it rise above mountains, houses or trees, you're guaranteed to have spectacular sights.

How to make a supermoon mosaic

Make a memory that lasts and capture a high-definition shot of a mega-Moon

The Moon is so close that getting a single, detailed image of all of its wonders is no mean feat. Instead, a lot of astronomers create a 'Moon mosaic' — a large image made up of several images stitched together.

One of the easiest ways to achieve this is by attaching a webcam to your telescope. You can even buy specialised Moon imaging equipment such as Meade's Lunar Planetary Imager, which plugs straight into your computer via USB. A location where you'll have a clear view of the Moon for several hours is also favourable. If you're successful, you could show off your creation by printing it onto a canvas.

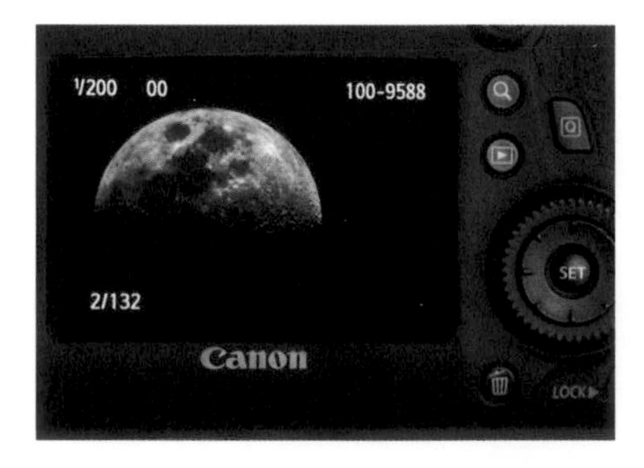

01 Find your location

Ideally you want a place far from street lighting that will give you an uninterrupted view of the Moon for several hours — the last thing you want is the Moon disappearing behind a tree halfway through your imaging! Make sure it is fully dark too — changing light conditions can be troublesome.

02 Set up your equipment

You'll need a telescope in order to capture the finer detail on the lunar surface. Set it up in the usual way and attach your chosen imaging equipment (either a DSLR camera, webcam or dedicated lunar imager). A red filter can also cut out some atmospheric disturbance, leading to sharper images.

03 Find the optimum settings

It is key that your images are in focus. To get the best focus, move your telescope to the edge of the Moon to get light and dark. Take a few test shots and zoom in to check for absolute focus. You should also experiment with exposure times to ensure that no part of the Moon is saturated.

04 Take the shots/video

Whether recording videos (AVI is best) or taking static shots, start at the top of the terminator and systematically work your way around the lunar surface. Splitting the Moon up into 20-30 sections is probably about right. It doesn't matter if the areas overlap a little.

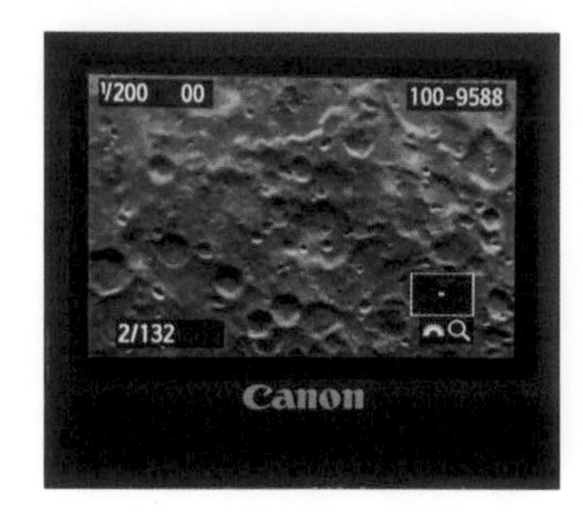

05 Process your mosaic

Use software like RegiStax to get the best frames from your videos. You can then use a piece of stitching software such as iMerge in order to build up your mosaic. Once you have your mosaic, polish it off by sharpening the contrast in photo editing software such as Photoshop.

Top supermoon targets

Whether you're using a telescope, binoculars or the naked eye, there are plenty of sights to be had on the lunar surface

Feature type:
Crater
Lunar Sea
Mountain range

Minimum optical aid:
Naked eye
Binoculars
Telescope

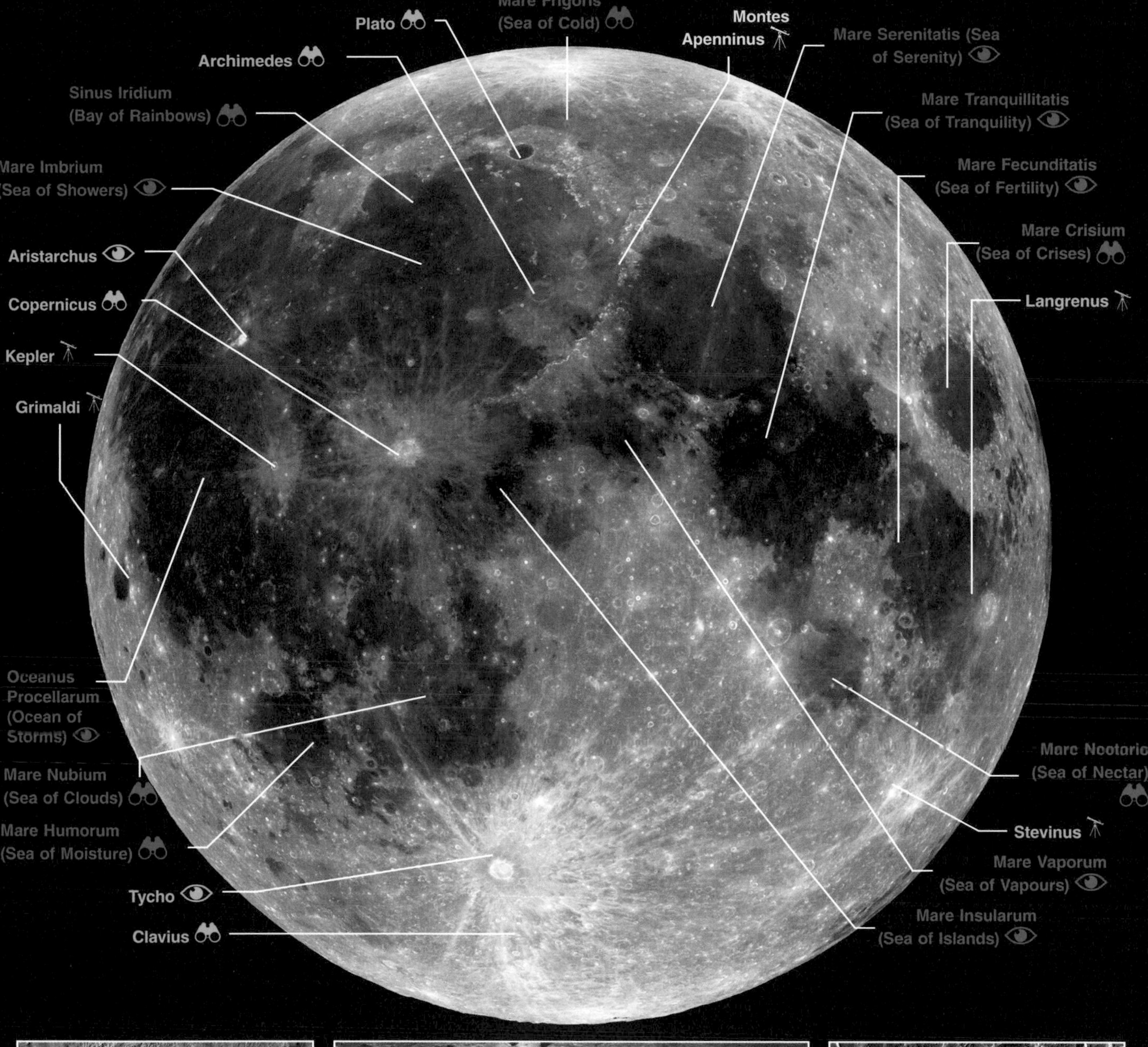

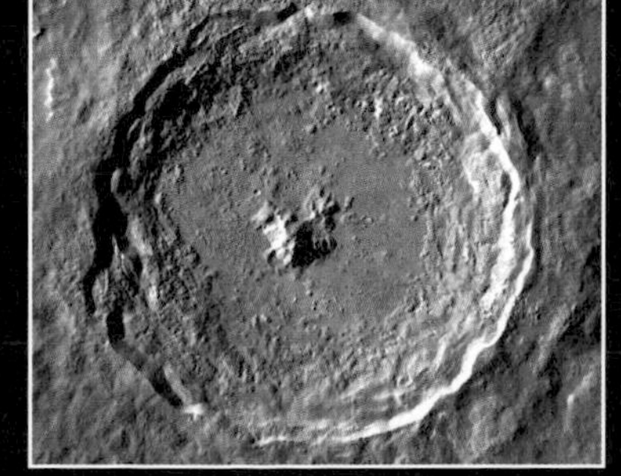
Tycho crater

Montes Apenninus

Calvius crater

© NASA

Secrets from the Far Side of the Moon

Often referred to as the dark side of the Moon, it's about time we went back – according to NASA's lunar astronauts and planetary scientists

Written by Nick Howes

Timeline

1959
Lunar 3 spacecraft takes the first photograph of the lunar far side

1962
NASA's Ranger 4 space probe becomes the first spacecraft to impact the far side of the Moon

1965
Soviet probe Zond 3 transits 25 pictures of the lunar far side

1966 to 1967
Lunar Orbiter program provides majority of coverage of lunar far side

1967
Prominent impact basin Mare Orientale photographed by Luna Orbiter 4

1968
Apollo 8 astronauts see the lunar far side for the first time

The Moon shows us its smiling Man in the Moon face every month, illuminated by the Sun to varying degrees over the course of its orbit around us. However, thanks to its orbital dynamics, we only ever get to see that one hemisphere from Earth. The other hemisphere — the 'far side' — is constantly concealed from us.

Cernan, Young and Stafford on board Apollo 10

Well, that's not strictly true. Libration, which is the gentle 'wobbling' of the Moon in the sky caused by changes in its position in tis elliptical (i.e. non-circular) orbit around Earth, means that we can catch glimpses of small slivers of the far side — we can actually see 59 per cent of the Moon's surface from Earth at different times of the year. But until the first space missions to the Moon flew around our natural satellite, what lay beyond on the far side was a mystery.

It's often mistakenly thought that the far side of the Moon is in darkness. Rather, it experiences day/night cycles just like the near side. When we see half of the Moon being illuminated by the Sun, giving it a half or crescent shape in the sky, half of the Moon on the far side is being illuminated at the same time. When the Moon is new, the far side is in full daylight instead. When the Moon is full, it's night-time on the far side. The reason we only see the one face is because of a phenomenon known as 'tidal locking'. The Moon rotates on its axis roughly once every 27 days, which is the same amount of time it takes to orbit the Earth. This means it is rotating at a rate that means we always see the same face, more or less, as it moves around Earth.

"There are two weeks of daylight and two weeks of night on every spot on the lunar surface," Charlie Duke, who was the Lunar Module pilot on the Apollo 16 mission, told All About Space. "It was early morning during the Moon day at the Apollo 16 landing site, which was called Descartes. We were the fifth mission to land on the Moon and I can say that it really is a dramatic place."

> **"There are two weeks of daylight and two weeks of night on every spot of the lunar surface. It's a dramatic place"** Charlie Duke

Apollo Capsule program manager George Low (left) alongside Wernher Von Braunm, the designer of the Saturn V

© NASA

Our first glimpse of the mysterious far side came early in the space race, courtesy of the Soviet Union's Luna 3 spacecraft almost 60 years ago. In 1959, barely two years after placing Sputnik 1 in orbit, Russian engineers managed to send the spacecraft, which was crude by today's standards, into orbit around the Moon and, for the first time, we got a good look at the mysterious far side.

Luna 3 took film images of the far side in total, which were photographically developed, fixed and dried on board — remember, this was long before multi-mega-pixel cameras. Ironically, the film used had been stolen from American spy balloons, as it had to be sturdy and radiation-hardened.

The spacecraft, using a combination of two camera systems — one wide field and one narrow-field but higher resolution — and a crude on-board scanner, could then send transmit the processed images, which were spot scanned from the photographs, back to the receiving station in the former Soviet Union. While only 17 of the 29 taken were transmitted successfully back to Earth, of which just six were considered good enough for publication, they proved to be a revelation.

Those 6 images covered 70 per cent of the Moon's far side and opened up a whole new perspective on the lunar surface. It was almost immediately evident that the dark patches that make the face of the Man in the Moon on the near side are almost completely absent on the far side. These dark patches are basaltic plains called mare, created by volcanic activity on the Moon billions of years ago.

"The computer told us that we were out of contact with the Earth and that we had loss of signal" Charlie Duke

Charlie Duke became the youngest person to walk on the Moon during Apollo 16

Instead, the far side was littered with craters, even more so than the near side, and some of those craters were the size of small countries. The Soviets started naming many of the features they were seeing for the first time, an act which caused some controversy in what was known as the height of the Cold War era.

We already had an inkling of one of those vast new craters, which is actually one of the very few mare on the far side. The subtlest hint of Mare Orientale, one of the largest impact craters known, seen on the limb of the Moon, had been known of since it's 'discovery' by Julius Franz in 1906 and can be seen during good librations when that portion of the Moon swings around towards us.

The view from Luna 3 showed how vast an impact crater Orientale was, resembling a bull's eye. It was almost 900 kilometres (560 miles) across — pretty much the length of the UK, give or take — and was caused by the impact of an asteroid impact, thought to be around 64 kilometres (40 miles) wide just under 4 billion years ago, and the resulting giant crater, termed an impact basin, was subsequently filled with volcanic lava.

In 1965, another Soviet mission, Zond 3, flew by the Moon with a far better camera than Luna 3 possessed and with the ability to conduct more detailed science observations, including spectroscopy. Zond 3 produced 23 very detailed photographs of the lunar far side, which enabled one of the first detailed maps of the entire lunar surface to be constructed.

In the meantime, NASA was progressing its Apollo Program at a phenomenal rate. Following the declaration by President Kennedy that the United States would place a man on the Moon and return him safely to the Earth by the end of the 1960s, by December 1968 NASA were ready to send three people — Frank Borman, Jim Lovell and Bill Anders — all the way around the Moon and back for the Apollo 8 mission. They became the first humans in history not only to escape from low Earth orbit, but to see the elusive far side.

This is how Lovell famously described the lunar surface: "The Moon is essentially grey, no colour, looks like plaster of Paris or sort of a greyish beach sand. We can see quite a bit of detail. There's not as much contrast between that and the surrounding craters. The craters are all rounded off. There's quite a few of them, some of them are newer. Many of them look like — especially the round ones — look like [they were] hit by meteorites or projectiles of some sort."

When their spacecraft flew around the far side of the Moon, the signal to Earth was cut off for around ten minutes. This loss of signal was a daunting time for the flight crew and mission control, alone and truly cut off from Earth, venturing where no human had ever gone before. As they came back around from the far side, a collective sigh of relief was breathed by many of the flight team at Mission Control in Houston. Charlie Duke describes what it was like to be flying over the far side of the Moon.

"The computer told us that we were out of contact with the Earth and that we had loss of signal," he says. "Then, all of a sudden, there was the sunrise, it was the most dramatic sunrise I've ever seen. In Earth orbit, you see the Sun's glow on the horizon or the planet's atmosphere and it gets brighter and brighter. The Moon is different though — there's instant sunlight with long shadows on the lunar surface. The far side of the Moon was very rough back there. I would not have wanted to land on the backside of the Moon."

After the success of Apollo 8, Apollo 9 went back into vital low Earth orbital testing of the lunar module, so the next astronauts to visit the far side were Gene Cernan, John Young and Tom Stafford on board Apollo 10 in May 1969, just two months before the historic landing of Apollo 11.

However, while flying over the far side of the Moon, the trio of astronauts

Two sides to the Moon

We can't see the far side from Earth, but the lunar faces are impressively different

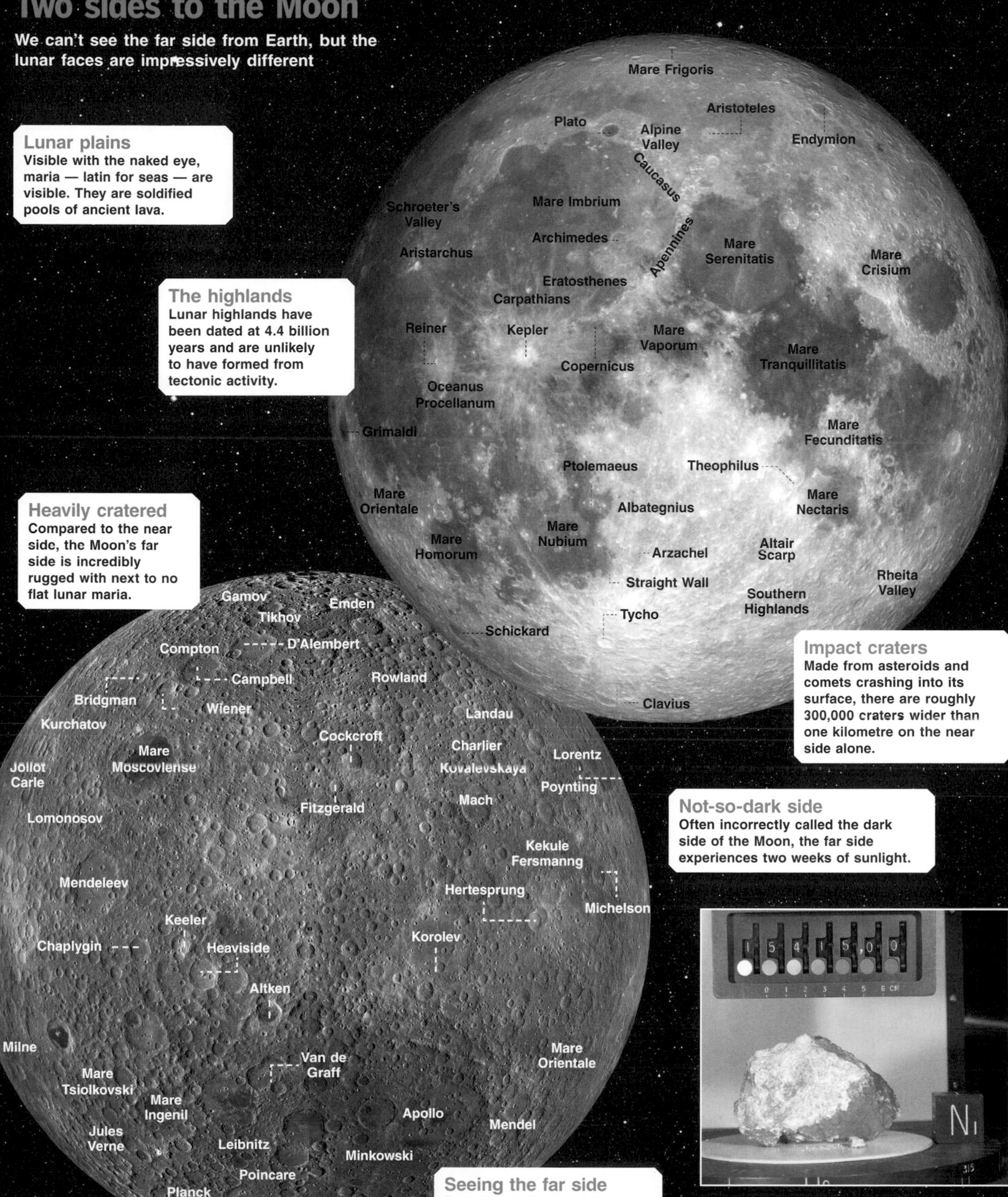

Lunar plains

Visible with the naked eye, maria — latin for seas — are visible. They are soldified pools of ancient lava.

The highlands

Lunar highlands have been dated at 4.4 billion years and are unlikely to have formed from tectonic activity.

Heavily cratered

Compared to the near side, the Moon's far side is incredibly rugged with next to no flat lunar maria.

Impact craters

Made from asteroids and comets crashing into its surface, there are roughly 300,000 craters wider than one kilometre on the near side alone.

Not-so-dark side

Often incorrectly called the dark side of the Moon, the far side experiences two weeks of sunlight.

Seeing the far side

From Earth, it's possible to observe a small proportion of the far side during libration - nine per cent is visible.

One of the hundreds of rocks collected during the Apollo missions, which are still being researched to this day. This being one of the most famous, the "Genesis Rock" from Apollo 15

encountered something strange, which in the last few years NASA has been forced to re-explain thanks to conspiracy theory documentaries airing on American television. The facts had been well known since the 1970s.

These strange events on Apollo 10 were manifest in the form of some very odd sounds. The radio system on board the Apollo spacecraft were crude by modern standards, though state of the art at the time. The command and lunar modules were relatively noisy environments according to most of the astronauts, with bumps and bangs combined with the whirring of fans and engine noise.

What the Apollo 10 crew heard through the radio systems baffled them. They described it as being almost like that made by an electronic instrument called a Theremin, often used in creepy science fiction B-movies of the 1950s and 60s, as well as on the Beach Boys song 'Good Vibrations'. Research has since proven that the sound was nothing more than an interference effect from those 1960s radio communications systems on board.

With the onset of the Moon landings, two astronauts would travel to the surface while a third remained onboard the command module to orbit the Moon alone, though all of them got chance to orbit the Moon and see the far side before landing. The solo orbital journeys of Michael Collins (Apollo 11), Dick Gordon (Apollo 12), Stuart Roosa (Apollo 14), Al Worden (Apollo 15), Ken Mattingly (Apollo 16) and Ron Evans (Apollo 17), who were the unsung heroes of the Apollo missions, are some of the bravest feats ever achieved by astronauts. They would spend days making quite detailed lunar

A replica of the Russian Luna 3 at the Museum of Aeronautics

Moon farms
A lunar farm would be stationed at the Lunar North Pole, allowing for eight hours of sunlight per day during the local summer by rotating crops in and out of the sunlight. Beneficial temperature, protection from radiation and insects needs for pollination would need to be artificially provided.

In an emergency
A short transit time of three days, which astronauts could improve on, allows emergency supplies to quickly reach a Moon colony from Earth or allow a crew to quickly leave the Moon and head back to our planet.

Lunar machines
With a round-trip communication delay to Earth being less than three seconds, it allows near-normal voice and video conversation and allows some kind of remote control of machines from our planet.

Transport on the Moon
The ability to transport cargo and people to and from modules and spacecraft would be essential on the Moon. Rovers are likely to be useful for terrain that is not too steep or hilly, while permanent railway systems could be used to link multiple bases and flying vehicles for-hard-to-reach areas.

observations from orbit, mapping features nobody had ever seen before.

Al Worden is often quoted as saying that his time alone was some of the best he had during the Apollo 15 mission.

"It was nice to be rid of those guys, as you can imagine, being stuck in something the size of a family car for over a week, it got pretty crowded up there. Once Dave [Scott] and Jim [Irwin] left, I felt like I had some real space to start to do my important work of mapping the lunar surface. But the far side, the views at certain times, when the Sun and the Earth are blocked out, are like nothing you could imagine. The sheer number of stars you see is incredible, it's like a sheet of white, and you know that every single one of them is a Sun in its own right."

A question often asked of the Apollo astronauts and flight teams is, why were all the missions just to the near side?

"We wanted to be in contact with the Earth, so we weren't able to land on the far side of the Moon," says Charlie Duke. Should something have gone wrong while the astronauts were on the surface, they would not have been able to communicate directly with Earth. This would not be such a problem today, as satellites could be put into lunar orbit to relay communications.

The far side is of growing interest to scientists, and potentially future planned human missions. Indeed, the possibilities for the far side of the Moon though are vast. For many decades the astronomical and scientific community have wanted to put radio telescopes and optical telescopes on the far side. Observatories on the far side would be shielded from not only man-made radio interference from Earth, but also the glare of daylight on our planet.The telescopes could be built inside craters to avoid solar radiation, and would provide us with an unprecedentedly clear insight deep into the far reaches of the universe.

We also have little true understanding of the processes that make the far side so vastly different in appearance to the near side. Why it is so scarred with impact craters and so lacking in volcanic mare is even more puzzling when you consider that when the Moon formed, it was much closer to Earth, and may not have necessarily been tidally locked at that time, meaning there would have been nothing special about the hemisphere we dub the far side.

Today, NASA's Lunar Reconnaissance Orbiter has mapped the near side and far side of the Moon in exquisite detail. When humans do eventually return to the Moon, the far side must be a goal for a landing. Understanding it will give us more insight into not only the Moon's past, but also perhaps the Moon's relationship with Earth our own past.

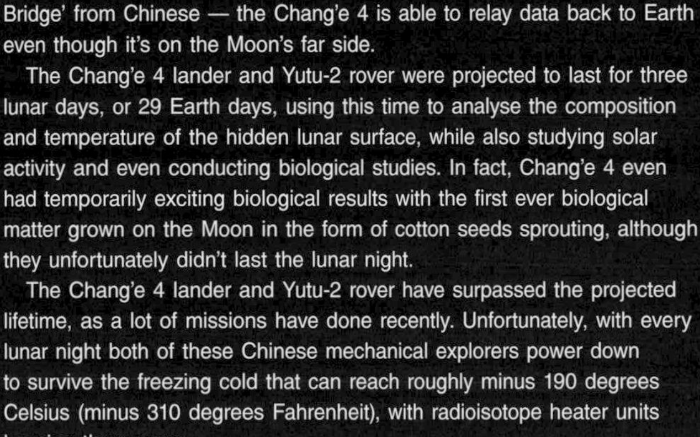

What the far side holds

China's Chang'e 4 mission explores the hidden side of the Moon

3 January 2019 saw the start of a historic mission as the China National Administration (CNSA) had successfully placed a lander and rover on the far side of the Moon. With the help of Queqiao — translating to 'Magpie Bridge' from Chinese — the Chang'e 4 is able to relay data back to Earth even though it's on the Moon's far side.

The Chang'e 4 lander and Yutu-2 rover were projected to last for three lunar days, or 29 Earth days, using this time to analyse the composition and temperature of the hidden lunar surface, while also studying solar activity and even conducting biological studies. In fact, Chang'e 4 even had temporarily exciting biological results with the first ever biological matter grown on the Moon in the form of cotton seeds sprouting, although they unfortunately didn't last the lunar night.

The Chang'e 4 lander and Yutu-2 rover have surpassed the projected lifetime, as a lot of missions have done recently. Unfortunately, with every lunar night both of these Chinese mechanical explorers power down to survive the freezing cold that can reach roughly minus 190 degrees Celsius (minus 310 degrees Fahrenheit), with radioisotope heater units keeping them warm.

© NASA

Who Owns the Moon?

Outside the confines of the Earth's atmosphere, who can stake claim on other bodies in the Solar System and beyond

Written by Christopher Newman

For the first time since the Apollo era, human space exploration is capturing the zeitgeist, with science fact and fiction blurring in a flurry of media excitement. Private companies are seeking investment to mine asteroids and groups of individuals are looking to raise money to embark on the colonisation of Mars. Never in human history has outer space seemed so accessible or replete with possibility.

But along with the formidable technical and engineering challenges facing such ventures are more prosaic issues regarding the rules and regulations for human space endeavour. As humanity moves away from low Earth orbit, the existing laws governing space activity will come under increasing scrutiny. It is a good time, then, to look at how the law regarding space activity has evolved and how it could respond to space mining and, beyond that, the colonisation of other planets. If space mining and colonisation provide the rich resources that they promise, a robust legal framework will be needed. Otherwise, the dream of space entrepreneurs could become mired in conflict and litigation.

At first glance, 25 November 2015 might not appear to be particularly noteworthy in respect of human space activity. Yet, the signing of the Space Resource Exploration and Utilization Act (part of the Commercial Space Launch Competitiveness Act) on this date means that it may well be one of the most significant days in the history of space mining. This piece of legislation, passed by the United States, is the first attempt by a nation at putting in place a legal framework for dealing with resources obtained by private companies from outer space mining.

When the laws governing space activity were drawn up in the first years of human space exploration, there was no conception that this type of activity would ever move beyond the pages of science fiction novels. In the late 1950s, under the umbrella of the United Nations, the Committee on the Peaceful Uses of Outer Space (COPUOS) established the rules that would lay down the basis of the regulation of space activity for the next 60 years.

The early formulations of space law, through UN General Assembly Resolutions, established a well-defined consensus preventing nations from claiming outer space for themselves, as well as the need for outer space to remain peaceful (if not entirely demilitarised) and specified that nations are responsible for their own space activities. The central trunk of space law is based on a treaty signed by over 100 members of the United Nations. Known colloquially as the Outer Space Treaty — or OST — of 1967

Astronaut Dale Gardner jokingly holds up a "For Sale" sign in space, but is it that easy to stake claim on an object in space?

Private companies such as ULA, SpaceX and Blue Origin are able to operate legally in space thanks to the Outer Space Treaty (the full title is somewhat lengthy: The Treaty on Principles Governing the Activities of States in the Exploration and Use of Outer Space Including the Moon and Other Celestial Bodies), it provides the basic framework in international law for all space activity and spawned a further four treaties.

Written at a time of tension between the USA and the Soviet Union, it is clear that those drafting the treaty wanted to prevent outer space becoming another theatre of conflict between the two dominant superpowers. Individual nations were made responsible for their space activities and retained responsibility for licensing jurisdiction and control of spacecraft and personnel. Significantly, the use of nuclear weapons in space was banned, which initially attracted the most attention and reflected the concerns of the time.

Addressing this desire for peaceful expansion in space, Article I of the Outer Space Treaty is aspirational in nature. It states that the exploration and use of outer space shall be carried out for the benefit and in the interests of all countries and shall be the province of all mankind. While this is difficult to translate into a specific legal duty, it does at least provide guidance as to the spirit in which space exploration should be undertaken. Article II of the Outer Space Treaty is, however, much more explicit and is crucial to understanding the current legal position in regards to space mining. The treaty says that outer space and celestial bodies are "not subject to national appropriation by claim of sovereignty, by means of use or occupation, or by any other means." In practical terms, the position in international law is that no nation can lay territorial claim to the Moon or any other celestial body. The Stars and Stripes left on the Moon by the Apollo astronauts is therefore purely symbolic — the United States, nor any other nation, can ever own the Moon.

"Unfortunately, your certificate saying you own a plot of land on the Moon isn't worth the paper it is written on"

Neither the Outer Space Treaty nor the other space treaties of the United Nations make any distinction between the Moon and other celestial bodies such as planets, asteroids, comets or even the Sun, seeking instead to make the whole of outer space a 'galactic commons'. Still, this has not stopped opportunistic businesses selling certificates promising lunar real estate or the naming of a star. You may even have bought one yourself, or given or received one as a gift. Unfortunately, your claim to a plot of land on the Moon or your right to name a star face two seemingly insurmountable problems.

First, without legal recognition of a national court (and as already explained, such recognition is expressly prohibited by the Outer Space Treaty) these claims cannot be enforced. The second problem involves your intention to take possession. Any legal claim for land must be accompanied by an intention to occupy that land, and — currently — there is no way for you or other people in possession of these certificates to actually live on their lunar acre or star. The International Institute of Space Law, writing in 2006, makes the legal position clear: no one owns the Moon, and it certificates that claiming a lunar acre or naming a star do not have any legal effect.

Given that the Outer Space Treaty was drafted at a time when space exploration required a superpower budget, it is unsurprising that it makes no mention of private space companies. The fact is, the Outer Space Treaty requires activities of "non-governmental entities in outer space to require authorisation and ongoing supervision" — in other words, companies such as SpaceX, Blue Origin and United Launch Alliance must obtain permission from the United States government before they can launch rockets into space. Companies in other countries must also do the same from their own nation's government.

In the same way, selling plots of lunar land or the naming rights of stars is also a commercial space activity and would require permission from the government to be legal (it's quite telling that no nation has ever licensed or approved such ownership schemes). The upshot of this is that, unfortunately, your certificate saying you own a plot of land on the Moon isn't worth the paper it is written on.

According to international law, no one can lay territorial claim to the Moon or any other celestial body, so the flag is purely symbolic

The Space Race

It was the growth of the rivalry between the United States and the USSR that saw the need for the Outer Space Treaty

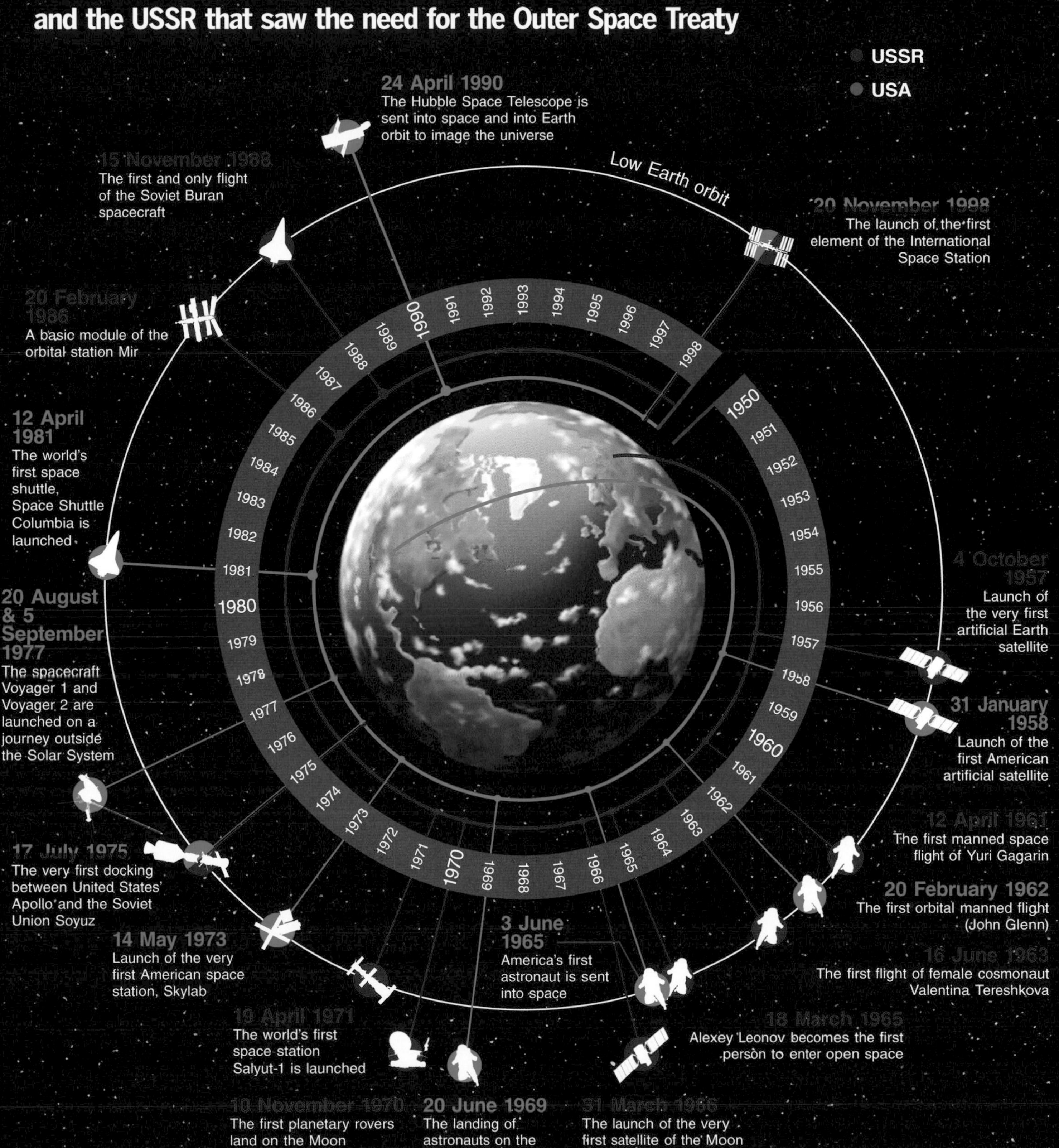

But when it comes to the mining of outer space, the legal position is significantly more ambiguous. Mining companies and commercial entities have no interest in laying claims of ownership or sovereignty on the celestial bodies; they desire merely to exploit the environment and extract minerals and other natural resources from the likes of asteroids.

Providing they are appropriately authorised by their state, the Outer Space Treaty allows private individuals and organisations to conduct activities in outer space and on celestial bodies. Where things become unclear is in regards to what can happen to the resources that are mined. The most crucial issue is whether companies can take ownership of the resources that they mine and, more importantly, make money from them. The trouble is, as we have already seen, taking possession of the resources would require legal recognition that the mining companies own those resources, which may run contrary to Article II of the Outer Space Treaty.

By 1979, the United Nations recognised that this might cause an issue and therefore the last of the big international space treaties, the Moon Agreement, was created. This stipulated that the Moon and other celestial bodies were "the common heritage of mankind" and any mining there should be administered by an international regime, mirroring the approaches taken in respect of the mining of minerals from the Earth's seabed. Yet there was no specific detail on how resources would be distributed.

So far, only 16 states have signed the Moon Agreement and, significantly, no countries with an active human space programme (USA, Russia, China) have signed up or even indicated broad approval. The Moon Agreement is, therefore, a failed treaty, and without any significant international support, it is unlikely to gain any traction.

The failure of the Moon Agreement leaves a significant gap in the regulation of space activity. Without it, the legality of mining currently depends on a positive interpretation of Article II of the Outer Space Treaty, with a private company subject to appropriate state oversight. In other words, a company may get away with mining resources in outer space so long as their activities are supervised by their government. The key word there is 'may': as long as things remain ambiguous it will do little to reassure potential investors in mining companies that their investment is safe! These investors would want to be able to enjoy the financial rewards from the mining, free from any form of legal challenge as to the ownership of any minerals. It is at this point that the US Government has stepped in with the Space Resource Exploration and Utilisation Act 2015. This allows US companies to claim ownership rights to the materials they mine in space and provide permission for them to transport and sell those resources.

Not everyone likes this new law, however. Although its proponents claim that it is consistent with the Outer Space Treaty in that the mining companies aren't claiming territory for themselves, what happens if two mining companies, perhaps operating in two different nations, claim the resources on the same asteroid, leading to dispute? In addition, some of these celestial bodies are of scientific importance and there's nothing in the new law to prevent these bodies from being ruined. Also, if you're not an American citizen or company, then you are excluded from recognition under the Act.

Nevertheless, these objections are currently just hypothetical — the real challenge won't come until mining companies begin work and start to put the law to the test. When humans start to move deeper into space, colonising as well as mining, we will need new laws and treaties that build on the Outer Space Treaty in order to successfully govern these colonies. These laws will no longer be about regulating specific activities, they will be the foundations of a new society. Social planning, dispute resolution and criminal justice will all need to be considered when thinking about a long-term future away from Earth. Even before those considerations are addressed in any longer-term approach to regulating space settlements, the Outer Space Treaty presents a number of key challenges for space colonists.

Colonising the Moon

Our lunar companion could serve as a stepping stone in surviving on other worlds in the Solar System

Launching rockets
A lunar base could serve as a site for launching rockets to Mars, using fuel that has been locally manufactured. It's easier to launch from the Moon than Earth since the gravity is lower.

Lunar machines
With a round-trip communication delay to Earth being less than three seconds, it allows near-normal voice and video conversation and allows some kind of remote control of machines from our planet.

Building an observatory
Making facilities for astronomical observations on the Moon from lunar materials would remove the need to launch building materials into space. The lunar soil can be mixed with carbon nanotubes to construct mirrors.

Close to home

Thanks to its proximity to Earth, at an average distance of 384,400km (238,855mi), the Moon is the most obvious place to colonise.

In an emergency

A short transit time of three days, which astronauts could improve on, allows emergency supplies to quickly reach a Moon colony from Earth or allow a crew to quickly leave the Moon and head back to our planet.

Lunar bases

Bases on the surface would need to be protected from radiation and micrometeoroids. Building a Moon base inside a crater would provide some shielding.

Moon farms

A lunar farm would be stationed at the lunar north pole, allowing for eight hours of sunlight per day during the local summer achieved by rotating crops in and out of the sunlight. Beneficial temperature, protection from radiation and the insects needed for pollination would need to be artificially provided.

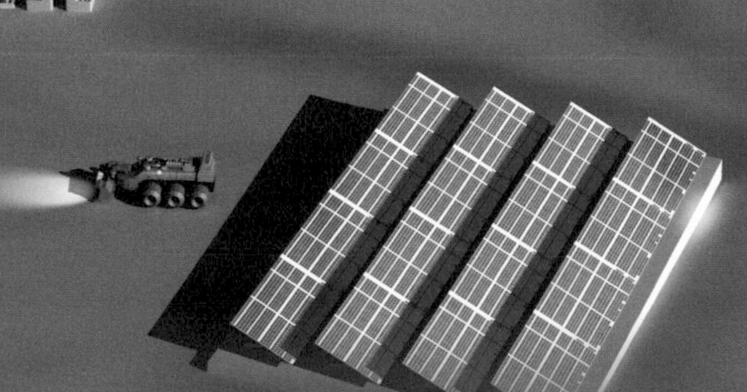

Transport on the Moon

The ability to transport cargo and people to and from modules and spacecraft would be essential on the Moon. Rovers are likely to be useful for terrain that is not too steep or hilly, while permanent railway systems could be used to link multiple bases and flying vehicles would be used for hard-to-reach areas.

Humans in low gravity

Colonising the Moon's surface means that we can find out how the human body responds to long periods of low gravity that's one-sixth that of the Earth's. We can then use this information to plan a viable a colony on Mars.

As with mining, the issue of sovereignty and ownership will feature highly in any discussion regarding the legal status of space colonies. As has already been stated, any mission, whether public or private, will need to be authorised by the government of the country from which it launches. A colony will, by definition, be occupying a celestial body and hence claiming it as their territory, which is in violation of the Outer Space Treaty. It is likely that, in the short term, a resolution similar to the one employed by the US in respect of space mining may well overcome this.

This does not, however, provide a long-term solution. Who is going to tell colonists that the celestial body, upon which they have lived for decades, is not really theirs? In addition, there are environmental issues, not recognised at the time of drafting the Outer Space Treaty, which will play a significant role in determining human conduct when building a home in outer space. Article IX of the Outer Space Treaty says that countries must conduct their space activities "so as to avoid their harmful contamination and also adverse changes in the environment of the Earth resulting from the introduction of extraterrestrial matter and, where necessary, [to] adopt appropriate measures for this purpose." However, critics say that these provisions are unduly interested in protecting the activities of states rather than the space environment. Even the contamination caused must be 'harmful' and the scope of this is not defined. Is leaving litter on Mars harmful in itself, or does it only become harmful when it destroys the environment of any hypothetical microbes that might live there?

The environmental impact of human space activity is only beginning to be felt in low Earth orbit, with swarms of space debris polluting the space lanes and creating hazards for spacecraft in orbit around our planet. Any legal framework in respect of space colonies will need to be mindful of the potential damage to the delicate space environment, but mining and ultimately colonisation will unquestionably have an environmental impact. Any new treaties will need to reflect this if humanity is to avoid polluting outer space.

Much of the discussion surrounding the legal issues of space exploration and exploitation is speculative and subject to advances in technology, simply because we haven't done much of it yet. So far, the Outer Space Treaty has remained the central trunk of international space law, guiding our behaviour in space. Even the American legislators in 2015 were still keen to emphasise compatibility and continuity with the Outer Space Treaty. For any student of space law, the first question to be addressed remains the same as it has been for the last 60 years: does the proposed activity offend against any provision of the Outer Space Treaty?

Ultimately, however, there is one key assumption that the legal framework governing space is based on: that life will not be detected on any of the celestial bodies in the near future. But even the discovery of microbial, non-terrestrial life forms will fundamentally affect the legal regulation of all space activities. Humans landing on the Moon acted as the catalyst for change leading to the drafting of the Outer Space Treaty. The discovery of extraterrestrial life would cause a fundamental rethink in the way that human space activity is conducted.

For example, would alien life forms have rights that would be recognised in human law? Could colonists keep them as pets, or bring them back to Earth where they could invade terrestrial environments in much the same way that transporting different species of plant to other countries can? It's impossible to answer these questions now; we must wait for these situations to become reality before they can be challenged in law. With the law regarding space mining and colonisation still being very far from settled, humanity's expanding frontier in space is going to be equally challenged by the expanding frontier in law.

Why we should mine asteroids

Asteroids provide natural resources to fuel the exploration of space and the prosperity on Earth as our population continues to grow

Water-rich asteroid

A single asteroid could produce enough fuel for every rocket launched throughout history

One single 500m (1,640ft) water-rich asteroid

An asteroid of this size would produce over £3.47 trillion ($5 trillion) worth of water for use in space. It currently costs about £13,892 ($20,000) to send a litre of water from Earth to deep space.

Uses of water in space

- Fuel for rockets
- Air to breathe
- Water to drink

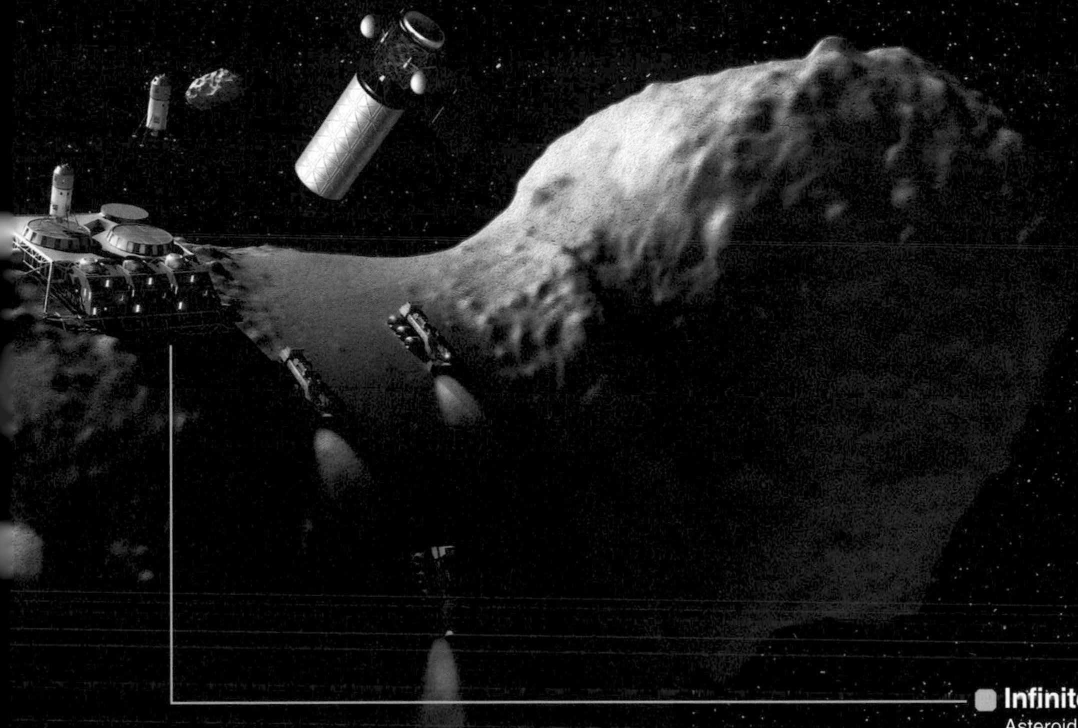

Infinitely rich

Asteroid mining will provide an almost infinite supply of platinum metals and water that can support us both on and off the Earth.

Plantinum-rich asteroid

This type of asteroid contains more platinum metals than we have currently mined from the Earth to date

One single 500m (1,640ft) platinum-rich asteroid

A 500m (1,640ft) platinum-rich asteroid is worth about £2 trillion ($2.9 trillion), which is more than our yearly output of platinum. Currently, 28 grams (one ounce) of platinum is valued at over £1,042 ($1,500).

Uses of platinum on Earth

- Reduces the cost of electronics

- Transport that requires electricity

- To create a greener Earth

© Adrian Mann; NASA; ESA; T.M. Brown (STScI); SpaceX, Blue Origin, ULA

How to...

Photograph the Moon

You too can take great-looking images of the lunar surface using your mobile device

© Nikos Koutoulas

You'll need:

- ✔ Smartphone
- ✔ Telescope
- ✔ Low-to-medium power eyepiece
- ✔ Moon filter that fits the eyepiece
- ✔ Optional phone bracket

Tips & tricks

Download software
Download and use an imaging app other than the standard one that's already built into the smartphone.

Take lots of images
Don't take just one or two photographs, take dozens. Each image will be slightly different and will help you to find the best software controls.

Experiment
Play around with the exposure controls of your software to see what difference it makes to your images.

You've probably seen photographs of the Moon in magazines and on the internet which reveal all of its craters, mountains and other features, but did you know that you can take similar quality pictures just with your smartphone camera?

Most smartphones have very high quality imaging devices built into them, as well as software that can give you some control over how they take the pictures, but many people just leave them on the automatic settings. However, if you have previously tried to hold your smartphone's camera up to a bright Moon in the night sky and attempted to take a picture, you've probably been disappointed with the result, as the images often come out showing a bright over-exposed disc and little else.

This is because the camera's software can't cope with the extreme contrast between the dark night sky and the very bright features of the Moon. This can be improved by using the exposure controls, which you may have in your camera app, but these can still be of limited help.

The best way to get really good images is by using your smartphone camera in combination with a telescope. It doesn't have to be a large telescope, either — even a modest amateur scope should give a reasonable image. It is possible to just hold your smartphone's camera over the eyepiece of a telescope and click the shutter, but you will soon find that this can be a very hit and miss technique. Just lining the camera up with the telescope's eyepiece can be quite tricky!

What really helps here is a smartphone adaptor, which fits around the telescope's eyepiece and gives a bracket to hold the smartphone nice and steady. It makes lining up the smartphone's camera over the eyepiece much easier.

Once you've done this you will be able to concentrate more easily on the phone's screen and make any necessary adjustments with positioning and exposure to create a good image.

If the Moon is nearly full when you are photographing it, you may find it is still too bright to get all of the surface features clearly in view. This is when you may need to use a Moon filter, which will drop the light levels and make the imaging process easier to control. For a small investment, you will be taking great images that will impress not only yourself, but your friends and family too. And, capturing the features of the Moon will be much easier, and much more enjoyable.

Shooting with an iPhone, Android or BlackBerry

Taking good pictures of the Moon is all about being methodical. Here's how to do it…

The important thing to do when taking images of the Moon through a telescope is to take your time and get each step right. This can make the difference between a mediocre image and a great one. If you are using a smartphone bracket, for example, make sure it's securely attached to the telescope or eyepiece. If the Moon is very bright on your chosen night of viewing, screw a Moon filter to the eyepiece. But if the Moon is only a thin crescent, you may not need one. If you follow the simple steps here, it should minimise any problems and maximise your chances of taking some really great pictures.

01 Find the Moon

Make sure your telescope is pointing at the Moon. You will probably need to adjust it later, but this will only be a small adjustment.

02 Assess the lunar phase

Is the Moon very bright? Is the Moon full? You may have to screw a Moon filter to the eyepiece, but do this before attaching the bracket.

03 Secure the bracket

Make sure the smartphone bracket is properly secured to the telescope and adjust it to bring the camera's aperture directly over the eyepiece.

04 Attach your smartphone

Once you are happy that everything is in place, put your smartphone in the bracket and start up the camera app of your choice.

05 Make adjustments

Make any adjustments necessary to get a good image of the Moon on your smartphone's screen, including moving the telescope if necessary.

06 Find the right exposure

Once the image of the Moon fills the screen of the smartphone, adjust the exposure settings of your camera software to get a good clean image. Make sure it isn't under or overexposed.

07 Start shooting!

Take lots of images and tweak the settings between each picture to see if you can get a really crisp image showing lots of detail. This will also help you to find the optimum settings for achieving a great photo.

How to...

Make a Lunar Analemma

Perhaps you've heard of a solar analemma, but did you know you can get the same type of image from the Moon? Find out how...

© Peta Jade; Getty Images

You'll need:

- ✔ DSLR camera
- ✔ Tripod
- ✔ Wide-angle lens
- ✔ Telephoto lens

Tips & tricks

Pick your spot
You'll need to return your camera to the same spot every day, so make sure that it is easy to get to.

Use a tripod
Use a sturdy tripod to reduce camera shake, especially if there is a wind or strong breeze at your location.

Use a DSLR camera
A DSLR enables you to change the lens easily from wide-angle to telephoto.

Check the time
Make a note of the time of your first shot and add 51 minutes each successive day during the lunar month.

Use a remote shutter
Create sharper images and reduce vibration and image blurring by using a remote shutter release.

An 'analemma' is a composite picture of usually the Sun, taken over the period of a year, which shows its shifting position in the sky as the seasons progress. It can take a year to create an image like this, so as you can imagine, it takes some dedication. However, you can create the same effect with the Moon in just 29.5 days, if you're lucky enough to have an entire month of clear night skies!

The image gives you an extended figure of eight pattern and can make a very attractive picture. As with the Sun, an analemma only exists as an abstract idea and as a compilation of images within one photograph, and this is the only way to see it.

The trick to creating a lunar analemma is to understand that the Moon returns to the same position in the night sky around 51 minutes later each day. Therefore, if you image the Moon around 51 minutes later each successive day over the course of one lunar month, or 29.5 days, it will trace out the figure of eight curve when the images are combined. This pattern is due to its elliptical orbit and its tilt.

You'll need a good, sturdy tripod and a way of marking its position, so that you can put it in the exact same spot each day. You'll also need some image processing software and a little skill in its use to get a good final image, especially as the thin crescent phases will need to be taken in daylight, or at least bright twilight. Use the wide-angle lens for the background shot and to get the positions of the Moon in each phase, as this will be the image upon which you will build your composite. An attractive building or mountain range can look good for the background, but remember to leave plenty of room in the sky in which to superimpose your lunar images.

The telephoto lens is used to get more detailed images of the Moon. During image processing these shots will be superimposed onto the background in exactly the right spots.

If you don't get a whole month of clear skies, you can always take a shot of the correct phase the following month and work that into your final image. Bearing this in mind, it may still take some time to build up your final analemma, but the result will certainly be worth it.

Building your composite

Create a dazzling image of the Moon's movements across the sky

Use a wide-angle lens for the background and Moon position shots and then swap to the telephoto lens or 'zoom in' and take some more detailed images of the Moon. You'll use your image processing software to superimpose these details onto the position of the Moon in your wide-angle image later on. The reason for this is to make the Moon look more 'real'. In the wide-angle shots it will seem very small and insignificant. You'll also need to vary the length of exposure to cope with the differing light conditions. Take a few images at various settings to increase your chances of getting a good shot each day.

01 Take a background image

Choose a good location and background for your image and take some well-composed shots. Be sure to leave plenty of sky in the images, as this will be filled with your analemma over time.

02 Keep track of the time

Keep an observation diary of the exact time of your first shot and note down the phase of the Moon each night. This will help keep track of your shots and assist while creating the composite later on.

03 Experiment with your settings

Take multiple images each night and vary the settings of the exposure time and ISO each time you photograph the Moon. This will ensure you get at least one good shot per night for your analemma.

04 Adjust your viewing time each day

Don't forget to add 51 minutes to the time of viewing for every successive day that you photograph the Moon during the lunar month. Your observation diary will help with keeping track of this.

05 Get more detail

Use a telephoto or zoom lens to get more detailed images of the Moon. These details will be superimposed onto the position of the Moon in your wide-angle shot using your software later on. Don't make the lunar disc too large though.

06 Edit your images

Once you've got all of your shots across the lunar month and selected the best ones, combine all of the images into a composite using computer software such as Photoshop. This will show up the analemma.

What if the Moon Exploded?

Find out how the Moon could cease to exist and if life on Earth could go on without our nearest neighbour in space

Written by Jonathan O'Callaghan

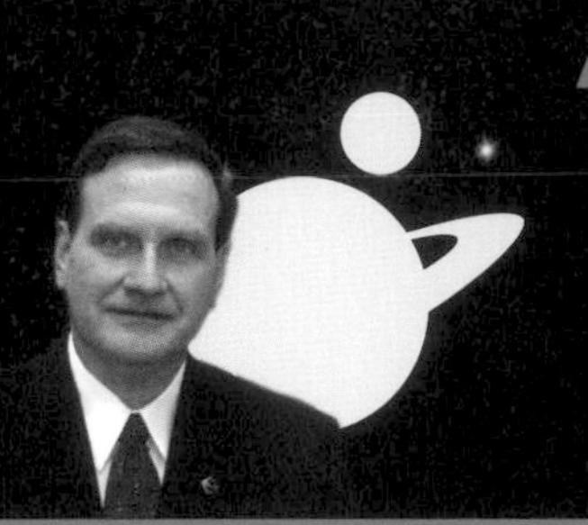

Interview Bio

Dr Paul Spudis

Dr Spudis is a lunar geologist for the Lunar and Planetary Institute at NASA. He is an advocate of using the Moon as an exploration port to the rest of the Solar System. In his prosperous time at NASA he has, among other projects, worked on the Lunar Reconnaissance Orbiter mission.

"First of all, how would the Moon explode? It might disintegrate due to a giant impact, such as a head-on collision of sufficient energy to break up the Moon. A condition of the produced debris would depend on the conditions of impact, a high-velocity impact would mean lots of vapour, which would then rapidly re-condense into billions of cooled droplets of glass. A low velocity collision would include large fragments tens to hundreds of kilometres in size and much of the debris would melt.

"The debris circling the Earth at lunar distance would not be stable, most of it would wander off into solar orbit, while some of it would stay in orbit around the Earth and re-accrete into a larger body. This would occur on time-scales of hundreds of thousands to millions of years. Without the mass of the Moon, lunar tides would no longer exist. However, the Earth and its oceans would still experience tides caused by the Sun. They would be of a much lower magnitude, but would still occur twice per day. There would be an immediate mass extinction of some organisms as intertidal species that depend on alternating periods of high and low tide would struggle to survive and adjust to the much smaller solar tides. The missing Moon would result in spin axis instabilities for the Earth and the obliquity (the angle from the path of the Sun in the sky) would oscillate wildly."

If the Moon exploded into small pieces, it's possible that over time they would form a ring around Earth, much like Saturn has today

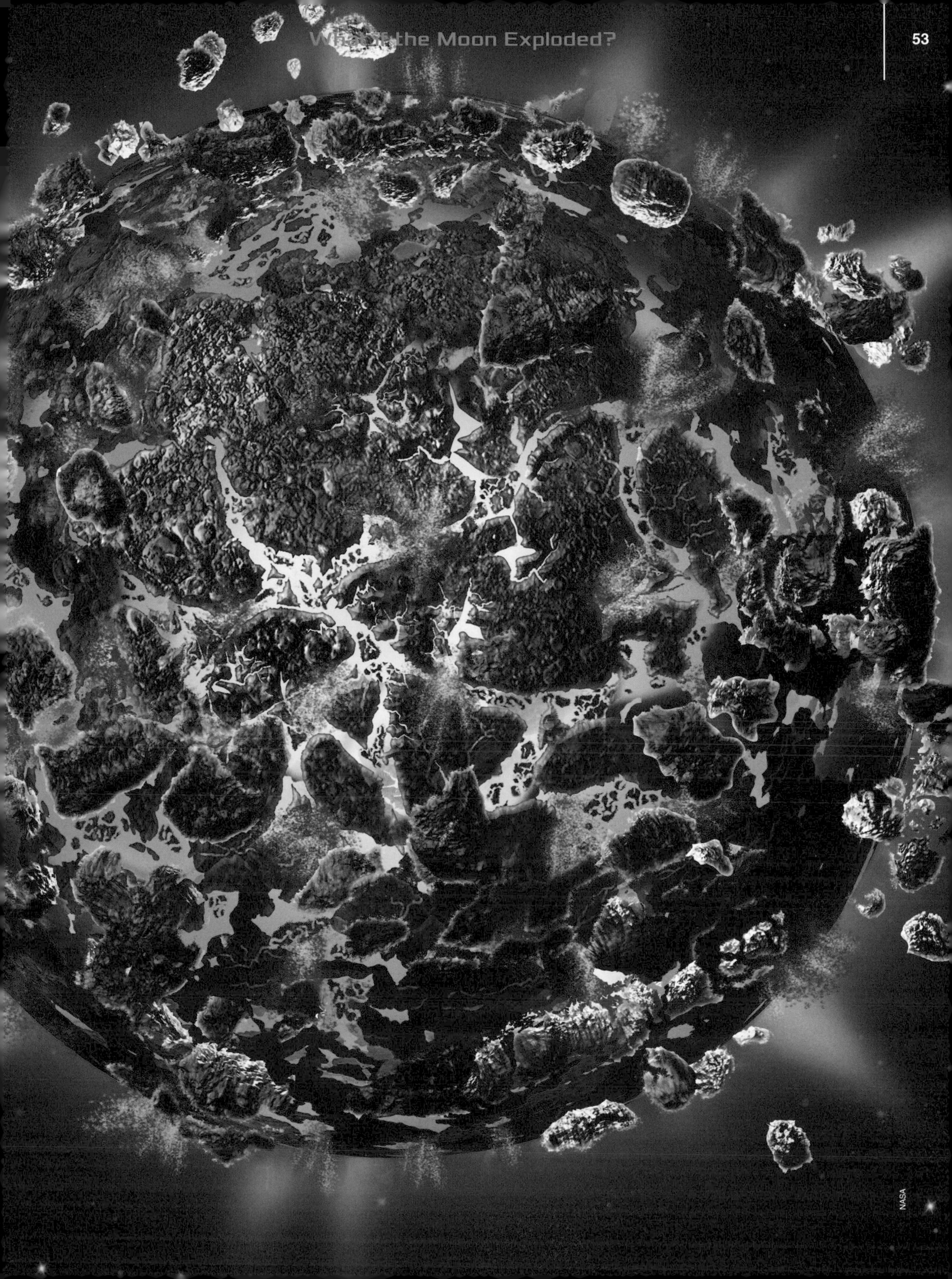

NASA

Moon Tour

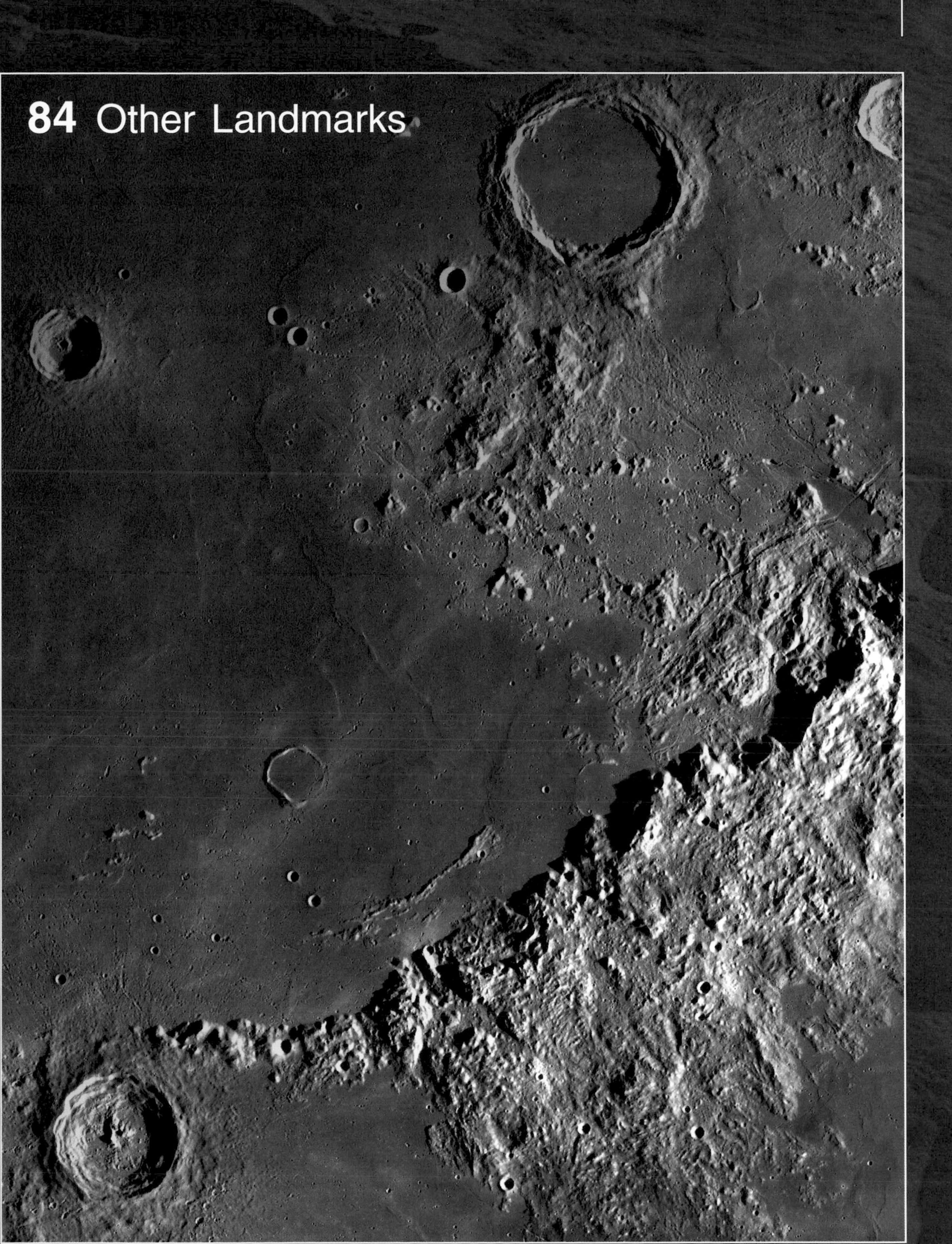

84 Other Landmarks

Craters of the Moon

Take a tour of the lunar craters on the Moon's surface

Tycho crater

Find one of the Moon's most famous craters, 48 years after it was immortalised in a classic science-fiction film

Whichever way you look at the full Moon — with the naked eye or through binoculars or a telescope — all you'll see are areas of light and dark. That's because at full Moon, with the Sun blazing overhead, features on the lunar surface cast no shadows and show no surface relief. The rugged lunar highlands are splashes of white, and the lower, lava-filled 'seas' are patches of blue-grey. But around full Moon is actually a great time for beginners to find what many people consider the most famous lunar crater: Tycho.

Tycho was named after the famous 16th-century Danish astronomer Tycho Brahe, a larger-than-life character best known for wearing a gold-silver nose after his own was cut off in a duel. But the crater's greatest claim to fame is that it featured in the sci-fi film *2001: A Space Odyssey*; it was where the enigmatic black Monolith was found by lunar researchers, triggering Dave Bowman's ill-fated mission to Jupiter.

Unlike some famous lunar features, finding Tycho is an easy task. Just look towards the bottom of the lunar disc when the Moon is full, rising mid-evening in the east, and you should see what looks like a bright spot. Binoculars or a small telescope will reveal the light spot has lots of narrow, bright lines radiating from it, some reaching to the top of the disc. This feature is Tycho, and those lines are rays of debris, which were created when the crater was born in a meteorite impact 108 million years ago.

The crater is 85 kilometres (53 miles) across and almost five kilometres (three miles) deep. The 'central peak' mountain that rears up from the centre of its hummocky, pitted floor is almost two kilometres (1.2 miles) high. The longest rays stretching away from it end more than 1,500 kilometres (932 miles) away, roughly the same distance as from London, UK, to Lisbon in Portugal. To see Tycho properly you'll have to wait until the crater is illuminated by the Sun at an angle, not from overhead, as its appearance will change dramatically.

When the Moon wanes and doesn't rise until 10pm, it looks like a classic crater — a pit with steep walls, a mountain peak jutting up out of its centre, and rays of debris shooting away from it on all sides. However, Tycho will be at its very best when the terminator of the waxing Moon silently sweeps over and past it.

With the morning Sun's slanting rays illuminating Tycho at a steep angle, even a small telescope will reveal a wealth of detail inside it. At high magnifications you'll see that the crater's inside walls slope gently and are terraced, with many clumps of material spread across the floor. The crater's central peak really stands out, too, though not as starkly as it did on the famous image taken by NASA's Lunar Reconnaissance Orbiter in June 2011. That landmark image revealed the mountain's slopes are streaked with rock-spills and strewn with huge stones, with an enormous single boulder, the size of Buckingham Palace, sitting right on its summit.

Tycho is one of those features that looks different every time you view it. When its eastern rim and western slopes are first kissed by sunlight, it looks like an empty eye socket staring back at you, but as the days pass, more details become visible. In those rare moments of perfect viewing, you'll see so much detail that you'll imagine you're flying over it.

Top tip!

The full Moon can be a dazzling sight through a telescope, so don't look at it for longer than a couple of minutes at a time. You can use a Moon filter to improve contrast and cut down any glare, which often washes out intricate surface details.

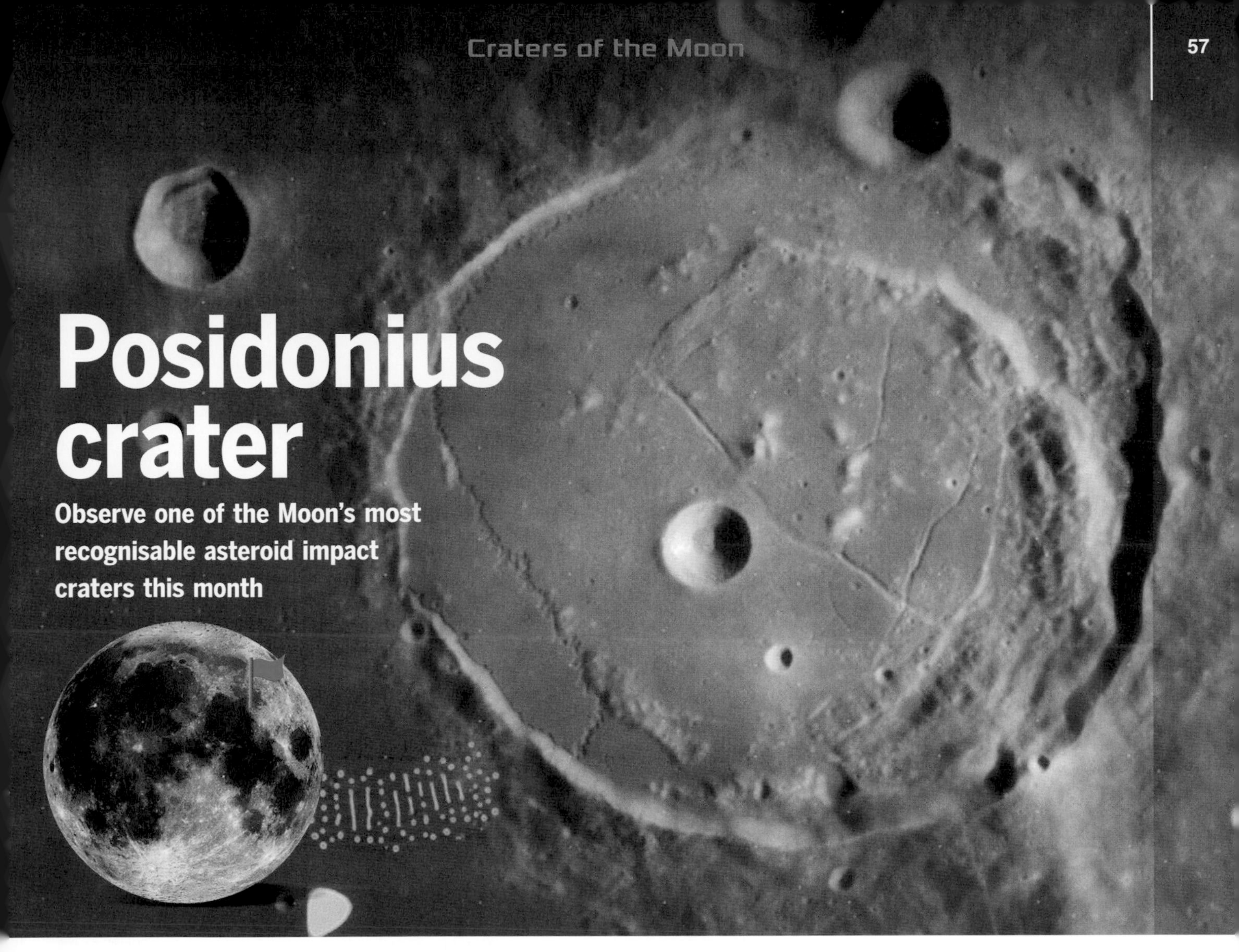

Posidonius crater

Observe one of the Moon's most recognisable asteroid impact craters this month

One of the Moon's most sublime and recognisable craters, Posidonius, can be viewed to good advantage when it is illuminated by a local evening lunar Sun. Two weeks later, the Moon lies low above the western horizon as the dusk skies are darkening, but vigilant telescopic observers may still be able to make out Posidonius near the lunar sunrise terminator, despite the Moon's low twilight altitude.

Measuring 95 kilometres (59 miles) across, Posidonius is a prominent crater that lies at the northeastern border of the large near-circular Mare Serenitatis (Sea of Serenity) and at the entrance to Lacus Somniorum (Lake of Dreams), an irregular, lava-filled plain that forms a short arc parallel to the northeastern shoreline of Mare Serenitatis. When illuminated, the Posidonius crater is easily identified through small optical instruments and even steadily mounted 7x50 binoculars. However, the complex nature of Posidonius is only revealed under good seeing conditions and through telescopes larger than 80mm at a magnification of higher than 50x.

Most noticeable is the fact that the southern rim of Posidonius is joined with the disintegrated crater, Chacornac. While both Posidonius and Chacornac are asteroid impact craters, the more eroded topography of Chacornac suggests that it predates the formation of Posidonius by several hundred million years. However, both craters are thought to have formed after the Serenitatis basin, as their outlines appear impressed over the basin's outline and both are filled with Serenitatis lava flows. Chacornac is likely to have been formed some time before the nearby Imbrium basin was formed, while Posidonius is thought to have been formed shortly after the Imbrium basin impact around 3.8 billion years ago, one of the final episodes of the Late Heavy Bombardment period, when the inner Solar System was a veritable cosmic shooting gallery of asteroids, comets and other debris from the formation of the planets.

Posidonius is one of the Moon's more unusual large craters, possessing a low but complete and clear-cut rim (north and northeastern parts of which are overlain by numerous smaller craters — Posidonius J, B and D). No traces of the initial external impact remain visible because of later lava flooding in Mare Serenitatis and Lacus Somniorum.

The interior of Posidonius presents a variety of complexity, though visibility is dependent on seeing conditions and the instrument/magnification used. A prominent, clear-cut, bowl-shaped crater, Posidonius A, sits slightly west of the floor centre at 12 kilometres (74 miles) wide — this is easily seen through small optical instruments. East of Posidonius A there are a number of small valleys; the linear rilles of Rimae Posidonius, the most prominent of which cuts 50 kilometres (31 miles) across the centre of the floor, and it is cut across itself at a right angle by a smaller rille. Another longer rille makes its way along the floor near the inner western wall. The best views of the Rimae Posidonius are to be obtained through a 150mm telescope at a high magnification.

A prominent curving ridge crosses the eastern part of Posidonius' floor — a huge block of crust appears to have slipped away from the main eastern wall. Chacornac appears much more irregular than Posidonius, with a crustal imprint that's obviously less fresh. Like Posidonius, it has an eastern central crater named Chacornac A (at five kilometres, or 3.1 miles, wide) and several rilles (Rimae Chacornac) that run northwest-southeast across its floor — easily visible through a 150mm telescope at high power and under a low illumination. Together, Posidonius and Chacornac present a highly interesting lunar sight when illuminated.

© NASA

Eratosthenes

Mid-sized, yet impressive, make the most of this well-defined asteroid impact that's set among the lunar mountains

Eratosthenes is a 59-kilometre- (37-mile-) wide crater located at the western end of the majestic curved sweep of the lunar Apennine Mountains (Montes Apenninus) on the southern shore of Mare Imbrium (Sea of Rains). This crater, set among a splendid mountain backdrop, is often overlooked by virtue of its proximity to the larger crater Copernicus, which lies less than 300 kilometres (186 miles) to the southwest and is often visible at the same time. However, Eratosthenes itself is such a spectacular feature when illuminated by a morning or evening Sun that its rugged topography is delightful through a mid-to-high magnification.

Eratosthenes is a typical mid-sized lunar impact crater, and its prominence is accentuated by its startling position at the southern end of the lunar Apennine Mountains. Formed by an asteroid collision that took place less than 2 billion years ago, Eratosthenes is around 1 billion years older than its near-neighbour Copernicus. As Eratosthenes first becomes illuminated by an early morning Sun (usually around a day after first quarter phase), the magnificent mountain arc preceding it forms a beautiful frame to the southern margin of Mare Imbrium.

When illuminated by a low morning or evening Sun, a prominent system of concentric and radial impact ridges shows up well around Eratosthenes. Owing to the crater's great age, much of the external sculpting around it has been hidden by later lava flooding by Mare Imbrium to its north and Sinus Aestuum (Bay of Billows) to its south, but a fair amount of impact topography remains to be seen through relatively small instruments under a low Sun.

Eratosthenes' floor displays a central mountain massif surrounded by a hummocky landscape that is stained with numerous dark spots and other variations in albedo. Unusually for such a large formation, Eratosthenes is barely visible under high angles of illumination — the crater's wall vanishes more or less completely, but its central peaks and dusky floor patches can be discerned at local lunar midday.

A century ago, Eratosthenes attained a certain amount of notoriety through the observations and opinions of astronomer William Henry Pickering, who observed the crater many times between 1919 and 1924. His vivid imagination convinced him that the dark spots on the crater's floor varied in intensity and, in addition, that the spots seemed to move around throughout the lunar day. Pickering, a fan of life on the Moon, speculated that the dark patches were vast swarms of lunar insects or herds of small animals constantly on the move, crawling or hopping around in search of sustenance. In reality, the spots on Eratosthenes' floor are composed of dusky surface material and are as motionless as its central peak. The patches do vary in their apparent tone, but so do countless other albedo features in response to the changing angle of sunlight.

When Eratosthenes is illuminated, you may just be able to locate the crater to the northeast of Copernicus, embedded within the impact rays of Copernicus, but the lack of shadows makes this a difficult identification for a first-time observer. The lunar sunset terminator then moves slowly from the east during the following days, and hints of topographic detail within Eratosthenes gradually reveal themselves as the shadows deepen. When the lunar morning terminator is encroaching upon Eratosthenes, the crater makes for a really splendid sight, preceding the mighty Copernicus on the sunrise terminator to its west.

A few weeks or so after this sees Eratosthenes illuminated, and the crater appears resplendent at the western end of the Montes Apenninus, while nearby Copernicus is also an extremely prominent feature. Low-medium and high magnification telescopic views are highly recommended; there are plenty of subtle features on the Moon in the area nearby to see, including chains of craters, domes and ridges.

Top tip!

When the lunar terminator is encroaching upon the Eratosthenes crater, it really makes for a splendid sight.

Cassini crater

Use autumn's crisp, clear nights to find a lunar crater named in honour of one of astronomy's most respected planetary observers

The Moon's 'celebrity' craters attract lots of attention because they are so easy to spot and are dramatic in an eyepiece. But the smaller, less dramatic craters — the ones without huge mountain peaks jabbing up from their centres or bright rays of debris surrounding them — can be just as fascinating if you take the time to get to know them.

One such crater is Cassini, which is perhaps the only noteworthy feature in the unremarkable lunar plain known as Palus Nebularum, just to the south of the famous Alpine Valley. As you might have guessed, Cassini was named after astronomer Giovanni Cassini, who, in 1675, was the first to observe the widest gap within Saturn's rings, later named the Cassini Division in his honour. In addition, Cassini discovered four of Saturn's major moons — Lapetus, Tethys, Rhea and Dione — and also observed markings on Mars, too. So it was no surprise when the Cassini probe, which orbiting Saturn for 13 years, was named after him.

The Cassini crater is just 57 kilometres (35 miles) wide with a lava-flooded floor. However, this floor is not flat; a pair of smaller craters (Cassini A and B) have been blasted out of its floor by impacts, and several much smaller craters spatter Cassini's interior. While B is unremarkable, A is quite an impressive crater in its own right — it's a deep, steep-sided oval pit some 15 kilometres (nine miles) wide, with an area of jumbled, hummocky terrain to its east, ploughed up by the impact which caused it.

Cassini's walls are narrow and low, without any of the complicated terracing or ledges seen in the walls of those 'celebrity' craters like Copernicus and Tycho, and some observers think they make it look like a ring dropped on the surface of the Moon. Using high magnification to view Cassini on a still night will show that the crater sits on top of a broad rampart of ejecta material, like a castle standing on top of a low hill.

So, when can you see it? When the Moon is almost full, Cassini can be hard to spot. Too small to see with the naked eye at any time, a pair of binoculars might just show it as a small, light ring to the lower right of the dark oval of Plato. But if you wait until the terminator sweeps towards it, its walls and craters will be much more obvious.

We lose sight of it a couple of days later, but Cassini reappears a week or so later when the Moon is just past first quarter and the first rays of lunar dawn are creeping over it. Then you have a few days to spot craters A and B nestling inside Cassini's walls, before overhead illumination from the Sun reduces the whole crater to a featureless bright stain on the Moon's disc.

Top tip!

The Moon can be a dazzling sight through a telescope, so don't look at it for longer than a couple of minutes at a time. Use a Moon filter to cut down any glare and improve contrast.

© NASA

Theophilus crater

Get to know one of the most striking but overlooked impacts on the lunar surface

As November slides into December, turbulent weather fronts often sweep across the country, scrubbing the atmosphere clean of haze to produce the first frosty nights of the year, with brittle air perfect for lunar observations. On these beautifully still nights, the Moon seems brighter and is a dazzling sight through binoculars or a telescope. It's always tempting to look at your favourite lunar features, but why not track down something you'll almost certainly have seen before but not spent time looking at properly?

Sandwiched between northern Sinus Asperitatis and the dark stain of Mare Nectaris to the southeast, Theophilus is one of the most striking craters on the Moon. Named after the 23rd pope of Alexandria, the crater is best seen four or five days after new Moon and just before last quarter, when it is close to the terminator. During full Moon, when the Sun is overhead, the crater's features are washed out and it looks more like a white smoke ring than a pit, but it is still easy to see.

Theophilus is a classic 'Tycho class' crater, closely resembling 'celebrity' craters like Copernicus, Eratosthenes and Tycho itself. At 100 kilometres (62 miles) wide and 3.2 kilometres (two miles) deep, it has a relatively flat, lava-flooded floor, pitted with many far smaller 'buckshot' craters. A quartet of 1.4-kilometre- (0.8-mile-) high mountainous peaks rise up from Theophilus' floor and viewed with a high magnification, they give the impression of a single peak hacked into four pieces. The tallest two peaks, on the western side, cast dramatic triangular shadows when the Sun's light strikes them at low angles. Like many other craters its size, Theophilus has a sharp rim and gently sloping walls. These are broken up with terraces, ledges and shelves on the eastern side, while much of the floor on the crater's western side has been covered by landslides.

Like the great crater Ptolemaeus, Theophilus is also one half of a double crater. It was blasted out of the Moon by an impact that obliterated the northeastern corner of another crater, Cyrillus. Although no major ejecta rays are visible stretching away from Theophilus — not even at full Moon — considerable amounts of material were scattered across the Moon when it formed. Some of this material was collected by the crew of the Apollo 16 mission, which landed in the Descartes Highlands region in April 1972.

So, when's the best time to see it? With the Moon approaching full, Theophilus is fully illuminated and shows virtually no surface relief. But a couple of week later, the crater becomes prominent as the Moon wanes. During the next evening the terminator rolls over the crater and it is plunged into darkness. It reappears two weeks later as the first rays of sunlight strike its walls. The Moon will be a beautiful slender crescent, low in the southwest after sunset.

When Theophilus looks its best, sunlight bathes its pitted floor and central peaks. At certain times, the Moon will be close to Mars in the sky, and the dark part of the Moon's face should glow with lovely grey-blue 'Earthshine', but don't let that distract you from enjoying Theophilus at its best! When the Moon approaches first quarter again, Theophilus will sink back into the lunar glare once more.

"When the Sun is overhead, the crater's features are washed out"

Langrenus crater

It's time to find one of the Moon's best hidden gems

The Moon has many 'celebrity' craters, like Copernicus, Tycho and Eratosthenes, which are big and bright enough to be obvious to the naked eye. However, these celebrities owe their fame to a stroke of good fortune: the bodies that blasted them out of the lunar surface millennia ago struck the face of the Moon pointing right at Earth. There are other craters just as big and interesting as Copernicus, but they are reduced to B- or C-List status because they were blasted out of areas not so well-placed for observation. Instead, we see them at an angle, foreshortened by the curve of the Moon's limb. Langrenus is one such crater.

A 137-kilometre- (85-mile-) wide, six-kilometre- (3.7-mile-) deep hole, punched into the Moon by a massive asteroid impact millennia ago, Langrenus would rival great Copernicus in beauty if it had been formed near the centre of the Moon's face. Sadly, it was blasted out of the eastern edge of Mare Fecunditatis, the ancient sea directly to the south of the dark eye socket of Mare Crisium, and so Langrenus' beauty and apparent size are both greatly diminished as it is almost on the Moon's limb.

Photographs taken by Apollo crews and lunar orbiters show Langrenus is very similar in appearance to Copernicus when viewed from above: it is a roughly circular crater, with shallow walls that are more than 20 kilometres (12.4 miles) wide and broken up into more than half a dozen slumped terraces and ledges. The walls are especially rugged and rippled on its western side. Stark mountains jut up out of the crater's floor with three-kilometre- (1.9-mile-) high peaks; these cast long, jagged shadows across the floor when sunlight hits them at a low angle after sunrise or before sunset. Beyond Langrenus' walls, out on the lunar plain, several rays of bright debris spread away westwards from the crater, but again their appearance is diminished by the angle of viewing.

One of the most striking things about Langrenus is the unusually high albedo — reflectivity, or brightness — of its floor. Its floor is very noticeably brighter than the surrounding terrain; it is more of a grey-white colour than the dark, ash-grey tones of the mare and landscape around it. This means that although the crater is reduced to an oval or lozenge shape by foreshortening, it is at least a bright one and, unlike some craters, it is easy to see whenever sunlight is falling on it.

Langrenus can be found as a small, bright oval shining near the western limb of the first quarter Moon, down at the 4 o'clock position on the Moon's face as darkness falls. Around full Moon, Langrenus will be a very noticeable bright mark beneath Mare Crisium through binoculars, looking like a bright smash pattern left in an icy puddle after a stone has been thrown onto it.

The best nights to see the crater are a few days later, when the Moon is starting to wane and the terminator begins to creep towards Langrenus from the west. With the Sun's rays slanting across the crater at a shallow angle, it will really stand out from the surface and look more like an actual crater. At this time, view it through your telescope with medium to high magnification, to see its central mountains and the shadows they cast across its floor.

However, after this, Langrenus is smothered by darkness and it doesn't emerge again until the Moon will be a beautiful thin crescent, low in the southwest after sunset. The crater will show some surface relief for the next few nights until all its shadows are washed away by the rising Sun.

Langrenus is also known as a hot spot of transient lunar phenomena — sudden brightenings that may be caused by releases of gas from beneath the crust — so make sure to keep an eye out.

© NASA

Aristarchus crater

Spend a night looking at the brightest major feature on the Moon

Top tip!

Don't be put off looking at the Moon when it's full, as it's a great time to see bright rays surrounding the youngest craters!

Many astronomers will tell you that the full Moon is the absolute worst phase of the whole lunar month to look at our nearest celestial neighbour; after all, all you can see are the Moon's major light and dark areas — its flat seas and rugged highlands, respectively. That's a little unfair. While it's true that you can see a lot of stark and fascinating surface relief on the Moon when it is a crescent, or first or last quarter, at full Moon you can see things not obvious at any other time.

For example, at full Moon it's much easier to appreciate how the mare are connected to each other (forming the classic "Man in the Moon") and it's also easier to appreciate the huge difference in albedo (reflectivity, or brightness) between the rugged, mountainous highlands and the dark, flatter lowlands. But the best thing about full Moon is that it allows us to see how incredibly far the rays of bright debris spray away from the youngest impact craters on the Moon, such as Copernicus and Tycho. And full Moon is the very best time to look at the brightest feature on the whole Moon — Aristarchus.

A quick Google search for "Aristarchus" will turn up countless amateur sketches and images of the crater; as it has interesting near neighbours, including Herodotus crater and the Schroter's Valley, the crater is a favourite of lunar observers and photographers alike. It has also been photographed by lunar explorers, both robotic and human. It was snapped during several Apollo missions, most notably Apollo 15, and more recently NASA's Lunar Reconnaissance Orbiter sent back stunning images of the insides of the crater's shiny walls, boulder-strewn floor and bright central peak. It helps that it is a favourite target for lunar observers and photographers alike.

Aristarchus is in one of the regions where observers claim to have seen transient lunar phenomena, or TLP, which is the term that Patrick Moore has coined for short-lived lights, colours or changes in appearance of the Moon's surface. Some lunar experts think TLP might be brief releases of gas coming up from beneath the Moon's surface, but it is an area of some debate, to say the least.

When the Moon is just a couple of days old, Aristarchus can't be seen as it is still deep in lunar night. It's not until the terminator finally sweeps over the crater that it emerges from the darkness. Then, up in the northwest quadrant of the disc at around the 10pm position, it will look like a small ding in a car windscreen through a small telescope. Three days later when the Moon is full, even just your naked eye will be able to pick out Aristarchus as a very obvious bright spot.

Binoculars will enhance its brightness considerably, turning the dot into a small tadpole or comma of striking silvery-white against the lunar disc, surrounded by a splash of fainter rays. Through a telescope at full Moon, Aristarchus is transformed into a fascinating splodge of white at the centre of an elaborate and beautiful web of silvery-white debris rays, as if someone has dropped a small pot of white paint on the Moon and it has burst open.

The crater remains visible as a white dab on the grey lunar surface until the terminator sweeps towards it once again. At this time, Aristarchus will look like a proper crater again, complete with terraced walls, flat floor and central peak. This will be the time to look at Aristarchus through the highest-powered eyepieces you have for your telescope, as you'll be able to see detail inside and around it that is not visible at any other time.

Make the most of the view on this evening and into the next morning because a day later the terminator will have rolled over the crater, plunging it into darkness again.

Plato crater

One of the Moon's greatest craters is an excellent lunar target

Known to pretty much all lunar observers, Plato is one of the Moon's most iconic and recognisable craters. Sunk deep below the level of the mighty lunar Alps, Plato is a near-circular depression around 100 kilometres (62 miles) in diameter, whose flat, dark floor lies more than two kilometres (1.2 miles) below its rugged mountain surroundings.

Because the Plato crater lies in the Moon's northern hemisphere, fairly close to the Moon's northern edge, a degree of foreshortening produced by the curvature of the Moon means that the crater doesn't appear circular from Earth; we see it as distinctly oval. In addition, libration — the apparent rocking motion of the Moon about its axis during the lunar month — affects the amount of foreshortening that does take place.

Although Plato was created by a substantial asteroid impact several billion years ago, no traces of the crater's original secondary impact formations can be observed. After formation, the crater's floor was quickly submerged by dark basaltic lava flows. Subsequent impacts and volcanic activity in the surrounding mountainous area and beyond masked the secondary impact formations around Plato — features that undoubtedly included prominent bright rays, chain craters and linear furrows. These features are not observable today; instead, we are left with an imperfect, though extremely prominent depression, set within the lunar Alps.

Following the solidification of the volcanic flows that spread across Plato's floor, a number of small impacts have made their presence known. These are so small as to be unresolvable through a small telescope, even at shallow angles of illumination. The five main craters found on Plato's floor range between 1.7 and 2.2 kilometres (1.1 and 1.4 miles) in diameter.

When illuminated by a low Sun at around the first or last quarter lunar phase, these craters can just be discerned through a 100mm telescope — their raised rims shine brightly against the dull tone of Plato's floor and they cast noticeable shadows. Under a high Sun, as at full Moon, Plato's floor craters appear as small bright dots that can be challenging to see through a 100mm telescope. Piles of material have slumped from the crater's inner wall, and a large triangular block of the inner western wall has broken away and slipped towards the floor, leaving a large dent in the crater's rim.

When illuminated by an early morning or late evening Sun, the shadows cast onto the floor by Plato's walls are fascinating to observe. Lit by the lunar morning Sun, the crater's western flanks remain in shadow, joined with the terminator, its inner western wall and the western part of its floor are bathed in streams of sunlight through low points in the crater's eastern rim. As the shadow cast by Plato's eastern rim recedes over the next day or two, the edge of the shadow projects into several points, shortening rapidly as the Sun climbs higher. A full Moon sees the crater fully illuminated, but its presence is clearly visible by virtue of its dark floor, set amid the bright nearby mountains.

Following the full Moon, shadows are cast onto Plato's floor by its eastern flanks. In the late lunar evening, the crater's eastern flanks are surrounded by the darkness of the terminator, while its inner eastern wall gleams as a bright crescent in the rays of the setting Sun as the floor darkens. One long shadow cast by a high part of the rim (to the north of the major landslide mentioned above) touches the base of the eastern wall. Several more long shadow fingers soon project across the crater floor, and the whole of Plato's interior, apart from the inner eastern wall, is plunged into darkness within just a few hours. The appearance of Plato's shadows constantly change from one lunation to the next, because of the effects of libration and the change in the direction of the Sun's illumination that it causes.

© NASA

Top tip!

Many craters, like Plato, are best observed during certain phases of the Moon. In particular, Plato is an excellent target during the first quarter Moon. If you choose to observe Plato during a full Moon, use a Moon filter to block out glare.

Aristoteles, Eudoxus and Montes Caucasus

Take the opportunity to observe two of the Moon's most prominent craters and mountain ranges

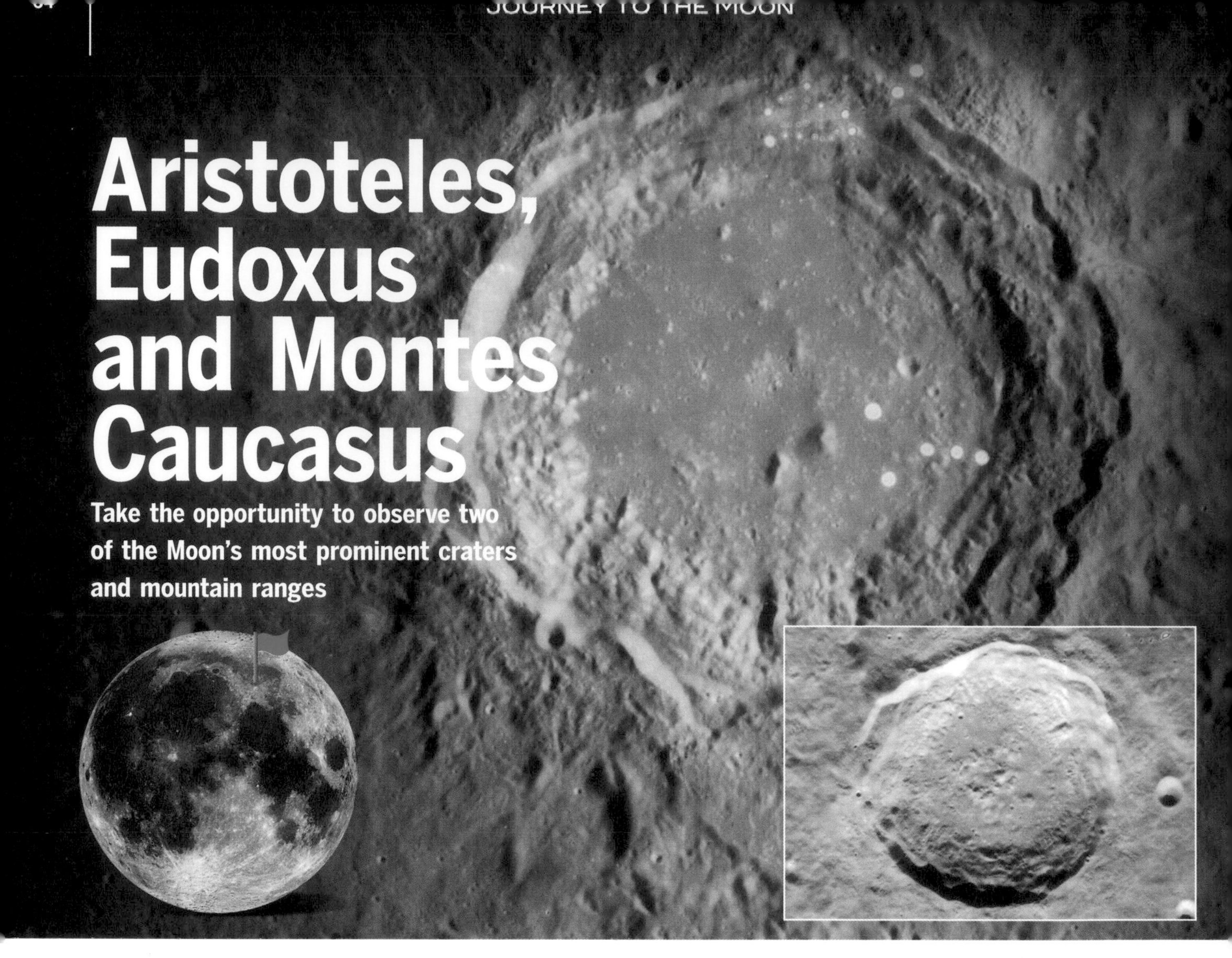

It's time to take a look at a notable visual partnership — one that is easily viewed through virtually any optical equipment. It involves two very prominent craters and a majestic mountain range that runs from north to south across several hundred kilometres of the Moon's upper northeastern quadrant, between the latitudes of 50 degrees and 30 degrees north.

Prominent craters Aristoteles (84 kilometres, or 52 miles) and Eudoxus (64 kilometres, or 40 miles) form a striking duo when seen through any telescope, particularly when they are illuminated by an early morning or late evening Sun. Indeed, the crater pair can be spotted whenever they are illuminated by the Sun — even through binoculars at full Moon — since both take the form of bright, well-defined rings when viewed under a high illumination.

Aristoteles, the larger and more impressive of the pair, has a slightly polygonal outline, along with broad inner walls that display some of the most extensive terracing within any lunar crater. Its floor is depressed below the mean level of the surrounding terrain, forming a flattish base around 40 kilometres (25 miles) in diameter, with two small mountain peaks protruding from the southern portion — features that are easily visible under the right illumination conditions using a relatively small telescope, specifically under 60mm.

The rim of Aristoteles is clear cut, displaying a scalloped effect that is seen in many other large impact craters of a similar size; such an effect is caused by large units of rock that have broken away from the crater wall and then have slid down it to some extent. Under a morning or evening illumination, a broad range of impact structuring can be discerned in Aristoteles' vicinity, taking the form of a mass of radial ridges that extend out from the rim for tens of kilometres. Buried within this structure on the crater's eastern wall is Mitchell (30 kilometres, or 19 miles), a crater that predates Aristoteles, providing a good example of a smaller crater that is overlapped by a larger one.

South of Aristoteles, the terrain becomes much rougher. At 100 kilometres (62 miles) to the south, Eudoxus makes an interesting neighbour — both Aristoteles and Eudoxus are easily encompassed within the field of view of a high magnification eyepiece. Although it superficially resembles Aristoteles, close examination of Eudoxus will show a number of differences. For example, Eudoxus' internal terracing is less ordered and its floor is somewhat blockier. The impact structuring around Eudoxus is less grand, partly because of the rougher nature of the surrounding terrain, with more concentric rather than radial structures being evident.

Aristoteles and Eudoxus are the bright jewels in a handle holding the south-pointing, highly-serrated 'dagger' of Montes Caucasus. This is an impressive mountain range that makes a north-south wedge more than 500 kilometres (311 miles) long, and separates eastern Mare Imbrium from the northwestern Mare Serenitatis. Peaks reach heights of 6,000 metres (19,685 feet) in places and the range appears particularly prominent when illuminated by (roughly) a first or last quarter Moon.

Top tip!

Prominent craters Aristoteles and Eudoxus form a striking duo when observed through a telescope, particularly when an early morning or late evening Sun illuminates them. A Moon filter will improve contrast, toning down any glare that often washes out intricate features of the lunar surface.

Endymion crater

Get to know one of the Moon's lesser known impact features

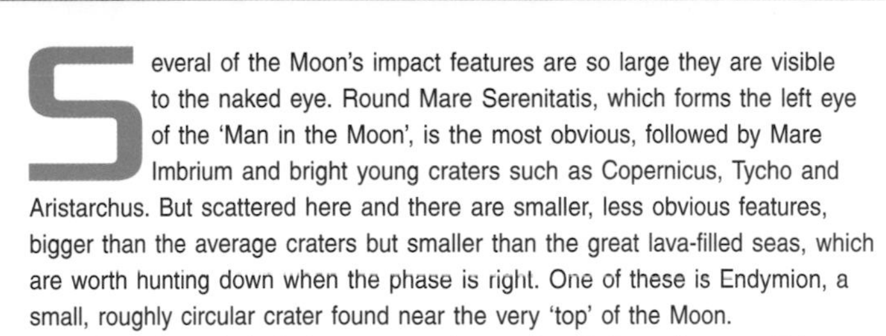

Several of the Moon's impact features are so large they are visible to the naked eye. Round Mare Serenitatis, which forms the left eye of the 'Man in the Moon', is the most obvious, followed by Mare Imbrium and bright young craters such as Copernicus, Tycho and Aristarchus. But scattered here and there are smaller, less obvious features, bigger than the average craters but smaller than the great lava-filled seas, which are worth hunting down when the phase is right. One of these is Endymion, a small, roughly circular crater found near the very 'top' of the Moon.

Endymion's mythological roots are rather confused; depending on which account you read, the crater was named after either a shepherd or a king. However, most of the stories agree that Endymion was so handsome and dreamy that Selene, the Titan goddess of the Moon, became so besotted with him that she begged Zeus to put him into an eternal sleep so that he would never age and she could drool over him forever.

If Endymion was farther away from the limb we would be able to see its true round shape, and it would look not unlike Ptolemaeus, or even a slightly lighter version of Plato. However, its close proximity to the edge of the Moon means that our view is greatly foreshortened. On the plus side, its location means that we can sometimes see it more clearly than others, thanks to the lunar libration, or its slight wobbling, which swings features close to (or even around) the limb into view.

Through a small telescope, Endymion looks like little more than a dark, oval spot, high above Mare Crisium, almost at the Moon's limb. However, larger telescopes reveal it to be a crater with a classic dark, flat floor. Images taken by lunar orbiters show that the floor is peppered with countless tiny craters, as if blasted by a shotgun, but they're so small it's unlikely you'll see them through your telescope. Up towards the top of the crater, you'll see the Endymion Triplet — a trio of craters arranged in a short line on the floor, pointing towards the crater's centre. They remind many observers of a mirror image of Orion's Belt.

Studies by lunar orbiters and ground-based observers show that Endymion is 122 kilometres (75.8 miles) across and 4.9 kilometres (3 miles) deep. Even from 384,400 kilometres (238,850 miles) away we can see that its walls are quite steep, and they can look very dramatic under the right conditions, when the crater is close to the terminator. Under high magnifications you'll see its far walls are terraced and slump in places.

When a beautiful crescent Moon hangs low in the west after sunset, Endymion is already fully illuminated, with the terminator — the line between night and day — having swept over it a little beforehand. By the time the Moon reaches first quarter, Endymion is losing definition, as the Sun's rays fall on it almost from overhead, shrinking the shadows cast by its walls.

By full Moon, the crater is reduced to a mere dull grey oval near the top of the lunar disc. But just a day later the terminator will be creeping towards it again, and its walls will be hit by more slanting rays of sunlight, briefly bringing them back into view for a mere day or so before the terminator rolls over the crater and plunges it into darkness again. Endymion will then be hidden from view until dawn breaks over the crater's walls and it reappears.

So why not try to drag your eyes away from your favourite lunar features and track down the Endymion crater? While it might not be the most dramatic or attractive feature to look out for on the Moon, once you've seen it, you'll definitely keep going back to it.

© NASA

Messier and Messier A

These two small craters are an intriguing sight through a telescope

Look at the Moon through a powerful telescope, a simple pair of binoculars or even just your naked eyes, and you can tell it had a violent past. Countless craters spatter its surface, each one a wound blasted out of the crust by the impact of a piece of rock or metal that came barrelling in from deep space. Most are so old that they're now just pits, empty eye sockets staring sightlessly from the Moon. But a few craters are young enough that they are still surrounded by systems of rays, bright lines of debris thrown out across the lunar surface when they were formed.

The largest ray systems, streaking away from the giant craters like Copernicus, Kepler and Tycho, are obvious to the naked eye on a clear night when the Moon is full. But here and there, dotted across the Moon, you can find smaller, less famous craters with smaller systems of rays, which are just as beautiful as their larger counterparts. One such crater is Messier A, a hole blasted out of Mare Fecunditatis (the Sea of Fertility) around 1 billion years ago.

Lying just south of the equator, Messier A is a mere nine kilometres (5.6 miles) across, which means you really need a telescope to see it, although it might just be glimpsed through a powerful pair of binoculars on a perfect night. It is actually one of a pair of craters, close to the crater Messier, which is more oblong in shape. However, Messier A is the more striking of the two because a pair of long, narrow debris rays jut away from it to the west, making it look like a comet.

In fact, its comet-like appearance is quite appropriate, because Messier A and its near neighbour were named after the French observer Charles Messier, the 18th-century comet hunter who compiled a catalogue of clusters, galaxies and nebulae in the sky so that he wouldn't mistake them for new comets. There has always been speculation about how Messier and Messier A were formed.

Some observers favour a 'double impact' scenario, in which a pair of objects hit the Moon simultaneously, creating a pair of craters. Another, and rather more romantic, theory is that a single body hit the Moon at such a shallow angle that after forming Messier it actually bounced back off the surface of the Moon, like one of the Dambusters' bouncing bombs bouncing off a reservoir — or, more recently, the Philae lander — and came down a second time, blasting Messier A out of the Moon and sending a huge spray of material away from it, forming the long, bright rays we see today. This actually makes sense: look at Messier through a telescope and it is strikingly elongated, which is what you would expect a crater caused by a low-angle, grazing impact to look like. So when can you see these intriguing craters?

When the Moon is a slim crescent, low in the west after sunset, Messier and Messier A will not become visible until the very young, crescent Moon is to the upper left of Mars and Mercury and the terminator sweeps over central Mare Fecunditatis, bathing the Messier pair in sunlight. The craters then remain visible until the terminator rolls back over them, plunging them into darkness once more. By then the Moon will be at its waning gibbous phase in the morning sky, to the upper right of Saturn.

When magnified through a larger telescope you'll easily be able to see the gap between Messier A's twin rays, and it should remind you of a classic comet tail, airbrushed on to the Moon's grey-white surface.

"Messier A is a mere nine kilometres across"

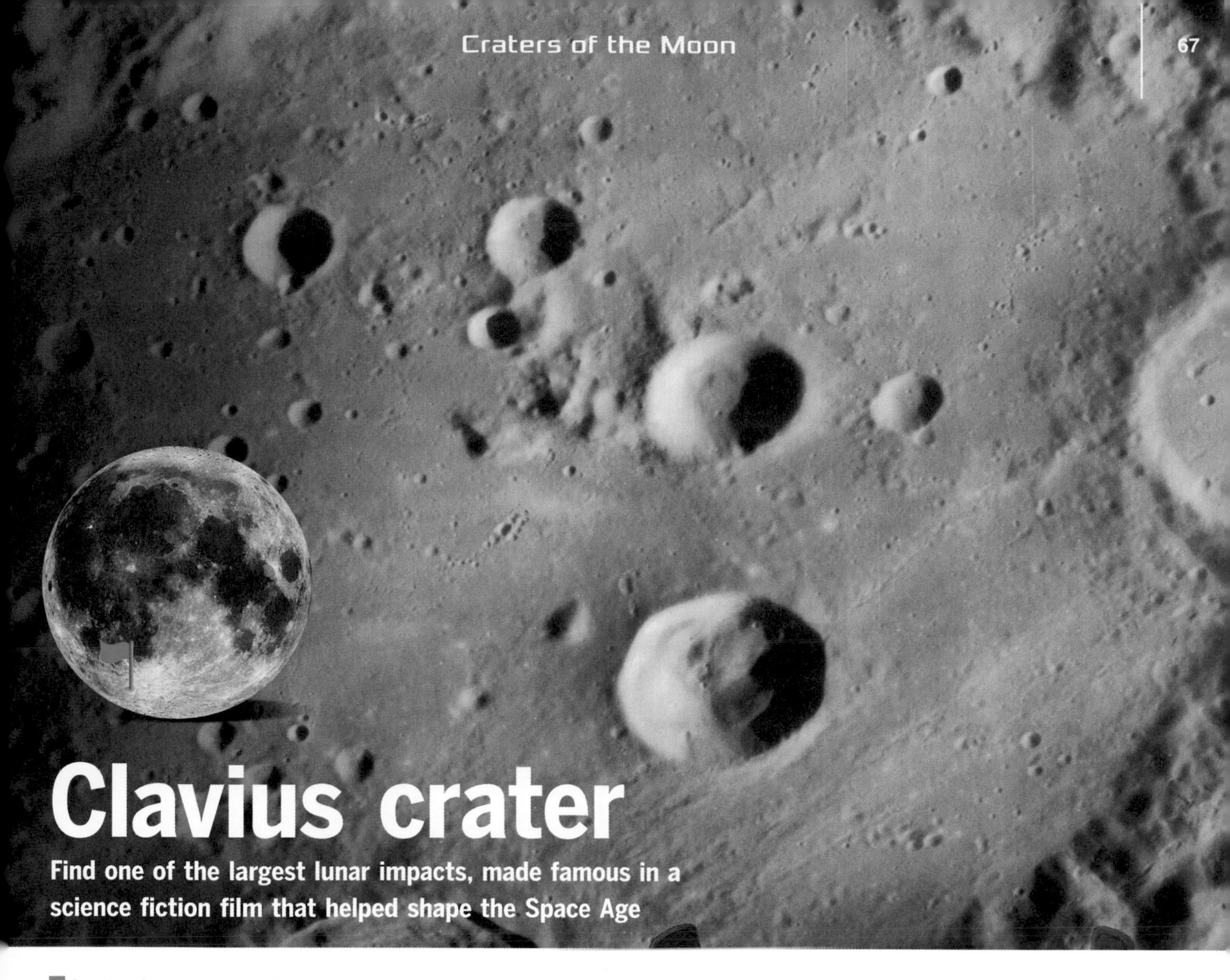

Clavius crater

Find one of the largest lunar impacts, made famous in a science fiction film that helped shape the Space Age

If you're a fan of the classic 1968 science fiction film *2001: A Space Odyssey*, this lunar 'celebrity feature' will need no introduction. No, not the bright-rayed crater Tycho, where the alien monolith was unearthed; Tycho is 430 kilometres (267 miles) to the north of our destination. We're going on a tour of the vast crater Clavius.

Clavius is so huge that it is easily visible to the naked eye when it lies near the terminator, appearing as an obvious notch between the areas of light and dark. Clavius' diameter of 225 kilometres (140 miles) makes it the third largest crater on the side of the Moon that faces Earth. Binoculars and small telescopes reveal a wealth of detail in and around the crater. The first time you see Clavius with even a little magnification, you will understand why this four-billion-year-old crater is a favourite object for many lunar observers.

Far away from any of the famous dark seas or historic Apollo landing sites, Clavius lies deep in the rugged and bright lunar southern highlands, 930 kilometres (578 miles) from the Moon's south pole, and is surrounded by countless other craters — large and small — on all sides. Its southerly latitude and proximity to the pole sadly means that we don't see Clavius at its majestic best; our view of it is cruelly foreshortened by the curvature of the Moon.

However, even a modest pair of binoculars will reveal a lot of fascinating detail in and around it. They will, for example, help you pick out a quartet of smaller craters nestled within Clavius' shallow walls, arranged in a curving chain which arcs upwards from the southern rim to the north. Very conveniently, the craters in the arc get progressively smaller in diameter the farther north they are, making this chain very useful as a test of an instrument's resolving power on a clear night. Binoculars will also show that those crater walls bear the scars of two huge ancient impacts: the oval, hummock-floored Rutherford Crater to the south and the rounder, flatter Proctor crater to the north are both easy to see at just 10x magnification.

It is through a telescope that Clavius truly comes to life. At higher magnification its floor appears spattered with countless craters, but you'll need really high magnification to see those. Clavius' northern walls show themselves as extensively terraced and rippled. It's as if a section of that wall collapsed at some point, sending a great tsunami of rock and dust sloshing towards the crater floor.

As is the case with all lunar features, you'll get your best views of Clavius when it is near or on the terminator — the line between night and day. Then, the sunlight will be slanting in at an angle, casting long shadows and highlighting the topography of the crater and its surroundings. When the Moon will have just passed first quarter, Clavius will reveal itself — look for the Moon low in the southeast after sunset, just to the left of the planet Saturn. Over the next few evenings Clavius will look better and better, quickly emerging from the gloom of the long lunar night. Fully illuminated by the low Sun its features will really stand out. By full Moon, Clavius will be past its best, pretty much washed out by the light of the Sun blazing directly above it. The crater vanishes from view when darkness crashes over it once more.

As you look at Clavius through your binoculars or telescope, just think back to that classic scene from *2001* when the shuttle is landing at Clavius Base. Back in 1968, as NASA prepared to land men on the Moon, it seemed possible — if not likely — that there would be such a facility in the year 2001, some 33 years in the future. Today, half a century later, we're probably no closer to building such a base as we were back in 1968 — a very sobering thought indeed.

© NASA

Top tip!

Copernicus looks the most impressive just after first quarter or just after last quarter Moon. At full Moon, it is just a bright spot.

Copernicus crater

Bow down before the Monarch of the Moon

Ask any group of Moon observers — absolute beginner or more experienced — to name their favourite lunar crater and the chances are most of them will nominate the same feature: Copernicus.

Copernicus has a nickname — 'The Monarch of the Moon' — and if you've only ever seen photographs of the crater but have never seen it with your own eyes, you would be forgiven for thinking that sounds a bit pompous. But, just as every amateur astronomer remembers their first view of Saturn and its rings through a telescope or their first display of the northern lights, they never forget the first time they saw Copernicus through a telescope. In fact, the crater is one of the first things a beginner will see through a pair of humble binoculars because it stands out so starkly on the Moon, even when viewed under 10x magnification through a shaking pair of binoculars on a chilly October night.

Copernicus was formed in fire and fury. Some time between 800 million and 1 billion years ago, the Moon collided with a large asteroid and the explosion caused by the impact blasted an enormous hole out of the area of the Moon we now know as Oceanus Procellarum, or the Ocean of Storms. The gaping pit left behind was almost four kilometres (2.5 miles) deep and more than 93 kilometres (58 miles) wide, which means its opposite walls are as far apart as London and Clacton-on-Sea.

However, the impact didn't just excavate a hole; it sent an enormous debris cloud of dust and rocks billowing up into the sky and across the lunar landscape. Some of that material fell back down to the Moon, spraying across it and leaving bright rays of ejecta splashed on the landscape. These debris rays are so bright and so long — the longest stretches for over 800 kilometres (497 miles), the distance between London and John O'Groats — that they can even be seen with the naked eye when the Moon is full, looking like white chalk lines drawn on the Moon's rugged grey-blue face.

The crater itself lies just south of Montes Carpatus (Carpathian Mountains) a short range of mountains on the southern shore of Mare Imbrium. Copernicus is roughly circular in shape, and its walls are terraced on all sides with a wide shelf dropped down on the western side, and landslides in several places. The crater's floor is quite flat, with a plain of ancient lava pocked here and there, and much smaller, much younger craterlets. In the centre of the crater a trio of mountains protrude from the lava plain, the tallest with a peak of one kilometre above the crater's floor.

Another of Copernicus' claims to fame is that it is relatively near one of the Apollo landing sites. In November 1969, Pete Conrad and Alan Bean guided the Apollo 12 lunar module Intrepid down to a landing site a little over 240 kilometres (150 miles) south of the huge crater. Their incredible precision landing placed them within walking distance of the unmanned Surveyor 3 probe, which had landed two and a half years earlier. The astronauts removed pieces of the probe and brought them to Earth for in-depth study.

Copernicus is fully illuminated when the Moon is a waning crescent glowing in the east before dawn. With the Sun's light slanting in at a low angle from the west, it will stand out from its surroundings and will be an impressive sight in a telescope. Magnified 100x or more it looks like an empty eye socket staring back at you from across space. However, it will only be visible until the terminator sweeps over, plunging the lunar surface into darkness. The crater will then be hidden from view until it emerges into the sunlight again, when the Moon is just past first quarter. As the nights pass the crater's appearance will change, its sharp outline and interior details blurring away as the Sun climbs higher in the lunar sky. When the Moon reaches full, Copernicus will be reduced to a flattened, light disc streaked by impressively bright rays.

Top tip!

To find Eddington crater, use Oceanus Procellarum as your guide. The impact is a bay within this lunar mare, with the more prominent crater Seleucus resting east-southeast of it.

Eddington crater

Challenge yourself by locating one of the trickiest impacts on the lunar surface

Most of the lunar features we are profiling are very easy to find, if not with the naked eye then at least through a pair of binoculars or a small telescope. They are large, bright craters, wide, dark seas or towering sunlit mountain ranges. This target, however, is going to provide you with much more of a challenge: it is small, tucked away almost on the Moon's limb and so tricky to spot you probably won't find it the first time you look.

Eddington crater is named in honour of British astronomer Sir Arthur Eddington, one of the first astronomers to figure out the nuclear processes that take place in the hearts of stars. Born in the picturesque Lake District town of Kendal, Eddington was a contemporary of Albert Einstein, with the great physicist calling Eddington a "genius". The author of many popular astronomy books, and a regular radio broadcaster, Eddington spent a lot of time doing what we refer to now as outreach — it could be said that he was the 1920s and 1930s version of Professor Brian Cox. He is perhaps best known for travelling to the island of Príncipe, off the coast of Africa, in May 1919 to observe a total solar eclipse. During totality, Eddington observed stars that popped into view around the eclipsed Sun. By measuring how much their positions appeared to have shifted as their light was bent around our star, Eddington helped prove Einstein's famous General Theory of Relativity to be accurate.

Eddington crater is an impressive 125 kilometres (77 miles) wide, and over 1 kilometre (0.6 miles) deep. If it was near the middle of the Moon's disc, it would be very easy to see, appearing larger than Copernicus and almost as large as Ptolemaeus. However, because it is very close to the lunar limb as seen from Earth, it is greatly foreshortened. It is also affected badly by libration, the rocking backwards and forwards of the Moon as it orbits the Earth: sometimes it is pushed towards us, and appears larger, other times it is pulled backwards and shrinks in size.

Whenever you look for Eddington crater, you'll see it looks more like an incomplete ring than a full circle, like most craters do — this is because it was flooded by a tsunami of molten lava many millennia ago. As the tidal wave rolled up Eddington's southern rim and spilled down into it, the crater was flooded, burying its southern walls and any central mountain peak it might once have had, leaving behind a horseshoe-shaped remnant with a flat, dark floor. Only a handful of small craters now pock that dark floor, the largest of which is Eddington P, that is a mere 12 kilometres (7.5 miles) across.

To see Eddington, you will almost certainly need a telescope, as only really powerful binoculars will spot it. Through a telescope's eyepiece, it will look like a vertically stretched, grey-white horseshoe against the darker lunar surface, very close to the Moon's curved limb. Eddington is not far from the much brighter and much more obvious rayed crater Aristarchus.

Being so close to the western limb, Eddington can only be seen when the Moon is at, or past, full. It is fully illuminated when the Moon rises in the evening sky, and as the month progresses and a waning lunar phase slips into the morning sky, the terminator will sweep relentlessly towards the crater, covering it. The impact will then remain unobservable until it is bathed in the Sun's rays once more.

> "This is because it was flooded by a tsunami of molten lava many millennia ago"

© NASA

Top tip!

Kepler will be at its best at one day past full, where the impact is fully illuminated and looking its best.

Kepler crater

Get to know one of the brightest features on the lunar surface

The Moon is a spectacular sight through any telescope — even a small one with low magnification — but not when it's full. Hearing that often surprises people; they think that 'bright equals best' when looking at something through a telescope, but the opposite is true. When the side of the Moon pointing towards the Earth is fully illuminated by sunlight, our planet's natural satellite is so bright that it's almost impossible to look at through a telescope eyepiece. If you can manage being dazzled by it for more than a few seconds, all you can see is a flat plate with grey and white patches splattered on it.

If you look at the full Moon through a modest pair of binoculars, or with just the naked eye, you will notice there are four very bright white spots with what appear to be bright lines streaking away from them — they look like the holes stones make in ice when thrown onto a frozen-over pond. These spots are in fact craters, which are relatively young when compared to the Moon's other impacts. One of these young craters is Kepler, which can be found on the western side of the Moon's face, below and forming a triangle with the equally youthful, but brighter and more obvious craters Copernicus and Aristarchus.

Kepler was formed around 1.1 billion years ago in what is known as the Copernican Era, when a huge asteroid barrelled in from deep space and slammed into the Moon's surface, blasting a 32-kilometre- (20-mile-) wide and 2.6-kilometre- (1.6-mile-) deep hole out of the lava plain of Oceanus Procellarum. Like its near neighbours, Kepler is notable because a huge amount of debris was thrown out of it, which sprayed up into the sky and splashed back down onto the lunar surface, leaving bright streaks and rays painted across it. The longest of Kepler's rays stretches for more than 300 kilometres (186 miles) across the landscape. There are more rays to the west of the crater, which suggests the asteroid which formed it came in from the east at an oblique angle, spraying debris away from the impact site.

Through a telescope at high magnification, Kepler is revealed to be a fascinating feature. It has a roughly polygonal shape, with outer walls that are sharply defined and smooth-sloped. The inside of the crater is more complicated as its inner slopes are terraced, leading down to a hummocky, crumpled floor dotted with smaller craters. Meanwhile, its central peak is neither very central or much of a peak, little more than a rough mound of material on the northwest part of the floor. To the north, large landslides have slumped down from the walls, and to the west a crescent-shaped ridge stands out starkly when the Sun's light strikes it at an angle.

Unlike some of the other major lunar features, no spacecraft — crewed or robotic — have yet landed close to Kepler, so its material has not been sampled. However, the crater has been photographed extensively over the years by many satellites and Apollo spacecraft, so we have very detailed imagery of it to study. In 2013, NASA's Lunar Reconnaissance Orbiter (LRO) flew over Kepler and its cameras took breathtakingly high-resolution images of debris flows, avalanches of dust and rocks snaking and slithering down its slopes, just like we have seen on Mars. The LRO images clearly show the debris flowed towards, around and even over obstacles in its path.

So, when can you see this fascinating feature? When the Moon is just one day past full, Kepler is fully illuminated and looking at its best straight away.

"Kepler is notable because a huge amount of debris was thrown out of it"

Menelaus crater

How to find the impact at the end of one of the longest rays on the Moon

Many well-known craters on the Moon are famous for being at the centre of complicated systems of rays. When they were formed, Copernicus, Tycho and Aristarchus all splashed bright rays of debris for huge distances across the lunar surface. Elsewhere on the Moon, smaller, lesser-known craters have systems of rays too, and one of them, Menelaus, is the source of possibly one of the Moon's longest and brightest.

Menelaus is a small crater, stretching just 27 kilometres (17 miles) across and barely three kilometres (1.9 miles) deep, but when the Moon's phase is just right it is easily visible in binoculars and small telescopes. Slightly oval in shape, it lies on the edge of the low Haemus Mountains, a curving mountain range that can be found on the southern shore of the Sea of Serenity.

Through binoculars the crater is quite unremarkable, little more than a sharp-edged pit blasted out of the ground, looking like the hole a nuclear warhead would leave behind in a cheesy science-fiction film. But no warhead created Menelaus; the crater was blasted out of the Moon by a chunk of solar system debris. The object came in at a shallow angle, resulting in the crater's oval shape. The impact scattered debris all around the crater, and created smaller secondary craters, too.

Menelaus also seems to be the source of a single, bright and very long ray that stretches away from it and crosses the whole of the Sea of Serenity, but even though the crater and the ray line up perfectly there is still a lot of debate about the relationship between the two. Known by some observers as 'the Bessell Ray', because it appears to be superimposed over the small crater Bessell, which lies to the northeast of Menelaus, this ray is a striking sight at full Moon and easily visible in binoculars. The mystery of the ray's origins probably won't be solved until robot probes or human explorers land on the Moon and can directly compare the ray's material to the that of Menelaus.

Through a small telescope, Menelaus itself shows some interesting detail. Its inner walls are subtly terraced, with multiple small, narrow ledges, and its floor is very hummocky, with many mounds and hills there to observe. If the Moon is in the right phase, a telescope will also allow you to see a network of fascinating cracks and troughs in the flat, grey plain of the Sea of Serenity just to the northeast of Menelaus, known as 'Rimae Menelaus' ('rimae' means gaps, or fissures). At low magnification they look like scratches on the Moon. Higher magnifications at the right phase of the Moon show a complicated network of criss-crossing features, small pits and troughs.

So, when is the best time to see this controversial crater and its fascinating neighbours this month? When the Moon will be full, so Menelaus will be fully illuminated by the Sun and will look like a bright spot on the southern end of Mare Serenitatis, at around the seven o'clock position. However, as the Sun sinks lower in the sky, the crater will be looking like a pit again, and then the terminator — the line between day and night — will start to pass across the crater, and it will have vanished from view the following day.

After that point, Menelaus crater will then be in total darkness for two long weeks before it reappears as the terminator approaches it once more. The crater will slowly come under full sunlight again and will remain visible until the Moon is full once more. The very best time to see the crater will be when sunlight is striking it at a low angle and making it stand out most impressively against the lunar surface.

© NASA

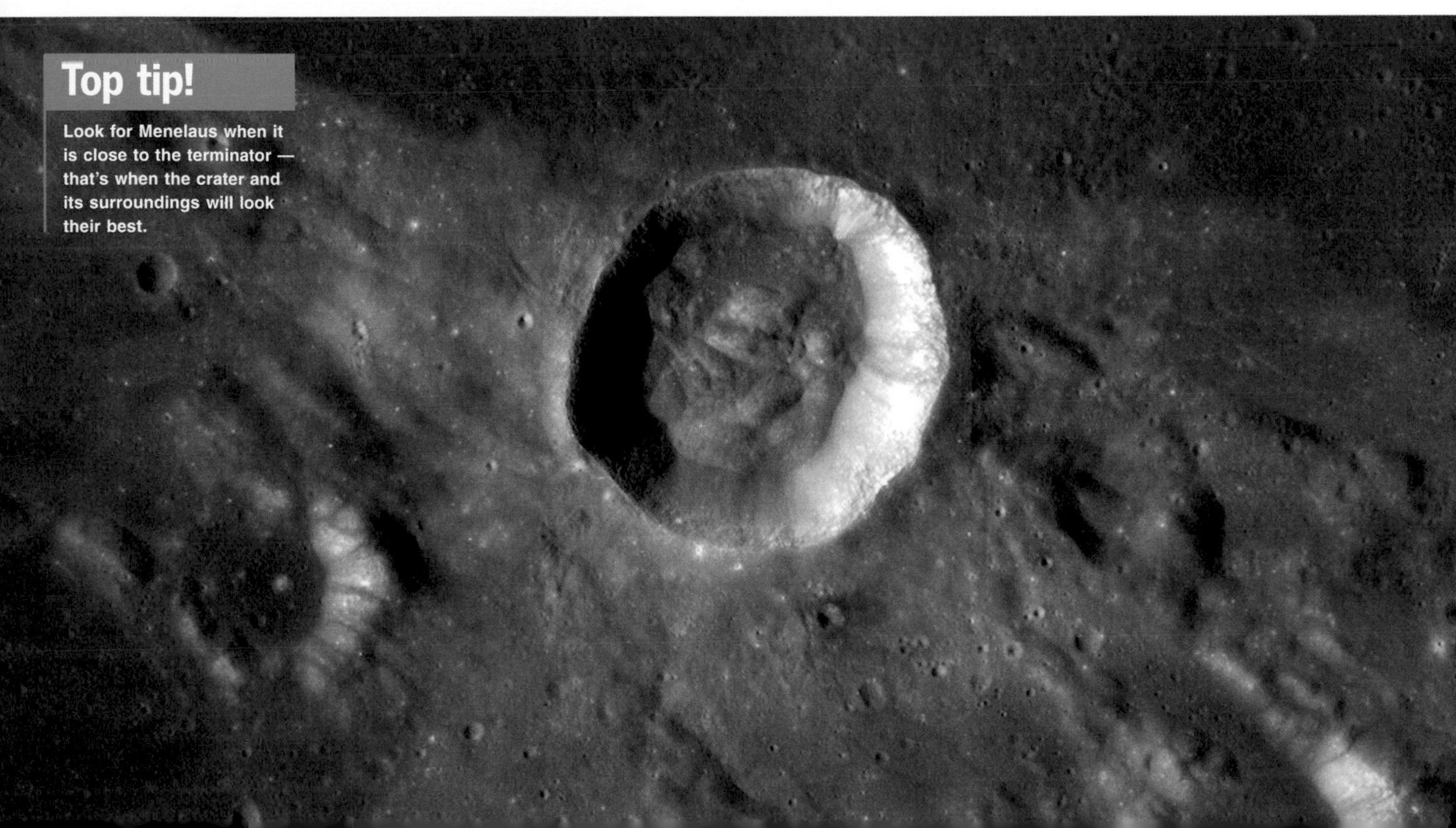

Top tip!

Look for Menelaus when it is close to the terminator — that's when the crater and its surroundings will look their best.

Schiller

Take a look at a fascinating lunar mystery

Everyone loves a good mystery, and astronomy is full to bursting with them. Astronomers are kept awake wondering: What caused the Big Bang? When did Saturn's rings form? Was there ever life on Mars? The Moon has many mysteries, too — how exactly was it formed? Will people ever live there permanently? Could it be terraformed to become a second Earth?

However, as puzzling as these questions are, the nature and history of the major features on the Moon's surface are mostly well understood. Thanks to centuries of telescopic study and decades of exploration with space probes and crewed landings, we know how its dark seas and impact craters formed, and why there are far fewer seas on the far side than on the Earth-facing side. But some features on the Moon are more puzzling. Here and there we can see bright debris rays following odd paths away from craters, and even strange, bright swirls on the dark maria. There are also a handful of craters that simply look strange. One of these is Schiller, a 179-kilometre- (111-mile-) long, 71-kilometre- (41-mile-) wide impact feature down in the south, not far from Tycho.

Through a telescope, Schiller looks more like a short, stubby valley than a typical round or oval crater. That's partly because being so far south and so close to the Moon's limb we see it at quite an oblique angle, so it appears foreshortened. Overhead views taken during the Apollo missions, or more recently by surveying orbiters, show Schiller really is an elongated feature. It's often described as 'lozenge-shaped' by experienced lunar observers, while others have compared it, rather less kindly, to a leech or a slug!

If you look at Schiller closely through a larger telescope using higher magnification on a night when the air is still, you will see quite a lot of detail. You'll be able to make out its rim, which is sharp and well defined, and its walls, which rise some four kilometres (2.4 miles) from its floor, are terraced with various shelves and ledges. There are two low ridges sticking up out of its floor on the western end, but the crater floor on the eastern end is very smooth and flat, with only a couple of craterlets pocking it. When the Sun is hitting Schiller at an angle it really is a fascinating sight, but at full Moon it blends into the background and becomes quite hard to find.

So, where's the great mystery? Well, we're not quite sure how Schiller was actually formed. At first glance, especially at low magnification, it's easy to think that it was created when a single chunk of space debris struck the Moon at a low angle, ploughing a long scar out of its surface. However, images taken from above suggest that Schiller really is not one, but two craters that formed at the same time, when multiple meteoroids slammed into the Moon almost simultaneously. How many? It was most likely a pair, but it could maybe have been as many as four meteoroids, according to some researchers. But Schiller isn't unique in this respect; the Orcus Patera crater on Mars looks very similar, and Venus has a crater called Graham, which also appears elongated.

However it was formed, Schiller is fascinating to look at. So when can you see it? It should start to pop into view as the terminator begins to sweep towards it — that's when its walls will start to cast shadows, making it stand out from the surface. The Moon will be in its last quarter phase and low in the sky before dawn, just to the left of the planet Mars.

The best time to see Schiller is when the Sun's rays are striking the crater at a low angle, making it much more obvious. However, when the terminator rolls over the crater, plunging it into darkness, it's lost from view until the cycle repeats again.

Top tip!

As with all mountains, valleys and craters on the Moon, you will be able to get your best views of this fascinating feature when it is close to the terminator.

Top tip!

Timocharis will be at its best as the lunar terminator approaches it.

Timocharis

Find a crater that had its mountain obliterated by an asteroid impact!

Even a casual glance at the Moon through a telescope, or just a humble pair of binoculars, is enough to tell you that it has been absolutely pummelled and pulverised in its distant past. Everywhere you look on the Moon you will see impact craters, the pits and holes left behind from a piece of space debris — an asteroid, a comet, or just one of the countless billions of chunks of rock that were left circling the Sun after it was born 5 billion years ago. Some are enormous and are surrounded by dramatic patterns of rays made of rock and dust that sprayed across the landscape when they were formed. Others are much smaller and less imposing, but there are many more of these than there are 'celebrity' craters. And some of these small craters have fascinating stories behind them, and incredible stories to tell.

One such crater is Timocharis, a small crater in Mare Imbrium, the Sea of Showers, the vast dark plain of ancient frozen lava that represents one of the eyes of the 'Man in the Moon'. Timocharis is just 34 kilometres (21 miles) across and three kilometres (1.8 miles) deep, so it is dwarfed by its near neighbours, mighty Copernicus to the southwest — 96 kilometres (60 miles) wide — and Eratosthenes — 59 kilometres (37 miles) wide — directly to its south, which it forms a tight triangle with.

Timocharis was given its name in 1651 by the lunar observer and cartographer Giovanni Riccioli. He named it in honour of the Greek astronomer and philosopher Timocharis, who, working from the famous Library of Alexandria, measured lunar star occultations very accurately and also observed Venus occulting a star.

Timocharis is a relatively young crater in lunar terms — we know this because it is not surrounded by any major rays or splashes of debris. When the Moon is full you can see some subtle rays spreading away from the crater, but they pale in comparison to those that can be found sploshing away from its aforementioned neighbours.

Through a telescope at high power, Timocharis is revealed to be a much more complicated feature than it first appears. The crater is roughly polygonal in shape, with sharply defined outer walls — a classic lunar crater in that sense — but the inner slopes of its walls are broken up into multiple terraces and ledges, all of which appear to have slumped down towards the floor. This means that when sunlight strikes Timocharis from a steep angle it can look like a target or bullseye on the lunar surface.

Look closely through your telescope eyepiece and you will see that Timocharis is home to a central mountain peak — or rather, it used to. Where a towering mountain once stood, there is now a small, deep crater because some time after Timocharis was formed, an asteroid came hurtling in from deep space, and like Robin Hood splitting a competitor's arrow, it slammed into the mountain at its centre, obliterating it and leaving behind a single crater that now stares out of the heart of Timocharis like an empty eye socket.

So, when can you see this fascinating but little-known crater? Timocharis is very hard to see when the terminator — the line between lunar night and day — is be almost on top of it. However, when the crater emerges into the sunlight, it will be easy to spot, looking like a small, round pit or hole above larger Eratosthenes. When the Moon reaches full, Timocharis will have blended into the background and will look just like a dark spot with a slightly lighter circular rim.

Timocharis will start to appear more prominent again when the terminator begins to approach it. When the terminator sweeps over the crater, plunging it into darkness, we will lose sight of it.

Little Timocharis might not be the most impressive crater on the Moon, but it's a reminder that every feature on the surface of Earth's satellite is worth looking at, and has its own story to tell. All we need to do is listen.

© NASA

Harpalus crater

This destination is the star of one of the most famous sci-fi films ever made

This crater was the uncredited star of a film that was released 18 years before *2001: A Space Odyssey* and stunned audiences with its realistic depiction of space exploration — *Destination Moon*. Of course, many lunar features and landmarks have been featured in science-fiction films and TV shows over the years. Some of the very first science-fiction films were silent movies that showed people visiting the Moon in glorified artillery shells fired from huge cannons, to meet either bizarre-looking aliens or dancing girls.

In the aforementioned classic 2001, astronauts excavated an enigmatic alien monolith from deep beneath the crater Clavius. In *Star Trek: First Contact* the USS Enterprise's time-travelling First Officer Riker waxes lyrical to warp drive inventor Zefram Cochrane about gazing up at the terraformed Moon in his century and seeing the lights of cities shining there, and Lake Armstrong too. The Moon has even been visited by Spongebob Squarepants!

But in 1950, the film *Destination Moon* was the first to attempt to show the Moon and the view from it realistically by featuring the stunning artwork and models of artist Chesley Bonestell. Many consider Bonestell to be the original — and still the best — 'space artist'. Although his lunar landscapes were much more dramatic and jagged than the ones the Apollo astronauts would gaze out on and explore decades later, they were still far more realistic than anything painted or shown on screen before. When *Destination Moon* came out, 19 years before Apollo landed on the Moon for real, it took audiences on a thrilling mission to a real crater that you can find for yourself on the Moon: Harpalus.

Sitting almost in the centre of the Sea of Cold, a long, narrow stain just above the beautiful crescent-shaped Sinus Iridum in the far northerly reaches of the Moon, Harpalus is physically a fairly small crater. Just 40 kilometres (25 miles) across and three kilometres (1.8 miles) deep, it is less than a third as wide as Copernicus and just one-eighth the size of Clavius.

Visually it is an unremarkable feature, not helped by the fact that its close proximity to the lunar north pole means that our view of it from here on Earth is greatly foreshortened, so it usually looks like more of an oval than a circular feature. However, the Moon's libration — the axial wobble it has which causes it to occasionally but regularly tilt features around its limb towards us and then away from us again, meaning we can sometimes see a little way 'around the edge' of the Moon — sometimes causes Harpalus to be tipped towards us, allowing us a much better view.

Photos taken from directly above by orbiting probes show Harpalus is roughly circular, with shallow, terraced walls, a trio of low mountains rising up from its hummocky floor and a system of rays spreading away from it. In this way it looks rather like a smaller version of Copernicus. If you look at Harpalus through a small telescope when it is at its best, you will be able to see right into it and will be able to make out details on its walls and floor, including a small crater there.

So when can you see this crater at its best? You'll need to wait until the terminator passes over it, allowing the Sun to shine on its raised rim — before that point, it will be fully hidden in shadow. When the crater is in full sunlight, it will bevery easy to see in a small telescope. It should even be visible — just —through a good pair of binoculars. The crater will remain fully illuminated until the terminator will return and begin to sweep back over it again, cutting off the sunlight.

Top tip!

Look for Harpalus when it is close to the terminator, the line between lunar night and day. It will stand out much more clearly here.

Albategnius

Gaze down into an ancient walled plain that puts some of the Moon's better-known features to shame

There are many features on the Moon that have acquired 'celebrity' status because they are genuinely impressive. Others aren't given the same credit or attention because they are overshadowed by their more impressive neighbours. One such feature is a walled plain called Albategnius, which can be found almost in the centre of the Moon's face as we see it from Earth.

If you're not familiar with this feature, that's no huge surprise. Albategnius is overshadowed by three huge and very famous features directly to its west: Ptolmaeus, Alphonsus and Arzachel. Linked together, and a striking sight in both a small pair of binoculars and a large telescope, these three craters are very popular observing targets, which is why poor Albategnius, just to their east, is usually overlooked. It's a shame, because it is a fascinating and rewarding feature.

Having said that, not everyone has ignored Albategnius. In 1610, Galileo observed it through his first telescopes and was so impressed by its appearance that he drew it, including it on his famous sketches of the Moon. In modern times Albategnius has been observed and photographed in rather greater detail by many lunar probes, and in 1972 the crew of Apollo 16 took some beautiful images of it as they orbited the Moon on the fifth and penultimate Apollo mission to land on the surface of our planet's fascinating natural satellite.

Although Albategnius looks like a large crater at first glance it is actually classed as a 'walled plain', so it is more like a small sea surrounded by high walls than a simple crater. It is approximately 130 kilometres (81 miles) across, surrounded by jagged walls that tower more than four kilometres (2.5 miles) above the lunar surface, and has lots of smaller craters spattered across its deep floor, around 40 of them.

To the north of the crater floor, a tightly clustered trio of these craters runs from west to east, itself presenting a very interesting sight through a small telescope. Albategnius also has another major crater inside its walls. Look down to the southwest and you'll see the crater Klein, a 43-kilometre (27-mile), steep-walled pit.

In common with many large craters, Albategnius has a central peak, a mountain that stabs up from its floor. Its summit is over 1.5 kilometres (0.9 miles) above the floor and topped with a small crater of its own.

Some observers think the mountain range has the shape of a ghost or an angel — you'll need to look at it through your own telescope to decide if you agree.

Albategnius' walls are, like those of most large craters on the Moon, very complicated features in their own right, with multiple terraces and ledges breaking them up in every direction and criss-crossed and cut into by valleys and gorges here and there.

So, when can you see this intriguing if overlooked feature for yourself? Like everything on the Moon, Albategnius is invisible in shadow. It doesn't emerge from the darkness until the terminator sweeps over it, surrendering it to the sunlight. A day later the plain will be fully visible, its high walls standing out starkly against the surface with the Sun's rays striking them at a low angle.

By the time the Moon is full, with the Sun blazing directly overhead, the plain will have been reduced to a mere dark patch. As the days pass and the terminator creeps back towards it from the east, Albategnius will become more and more prominent again, until it is swallowed up by the darkness once more and is lost from our view.

So why not take a look at Albategnius for yourself? True, there are larger and more dramatic features around it, but if you can drag your eyes away from those and dare to stray from the well-worn path that you usually follow across the Moon, you'll no doubt find Albategnius to be an incredibly rewarding 'off the beaten track' destination.

© NASA

Lunar Maria

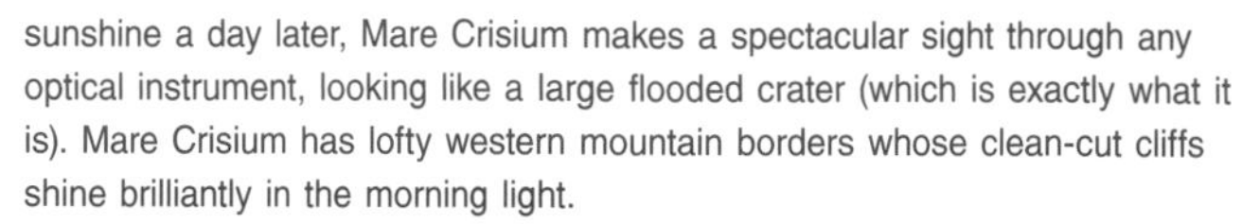

Explore the Moon's seas and how they got their names

Mare Crisium

Fully exposed in morning sunshine, Mare Crisium offers spectacular bright views

Despite their name, the prominent lunar features known as 'seas' ('maria' in Latin) aren't stretches of liquid water. They're vast pools of dark lava that flooded low-lying areas (mainly impact basins) several billion years ago. These lava flows have long since solidified. It was only in ancient times — long before telescopes were first invented — that these dark patches, so clearly visible with the unaided eye, were speculated to be marine regions. But for more than 400 years we have known that the Moon's seas are decidedly solid areas: no waves lap at their shores and no wind billows across their expanses.

Mare Crisium (the Sea of Crises) is the most 'self-contained' sea on the Moon's nearside. Viewed from above, it is markedly oval; measuring 570 by 450 kilometres (354 by 280 miles), its longest axis is oriented east-west. Viewed from Earth, Mare Crisium's position near the Moon's eastern edge causes its east-west axis to be 'squashed' — foreshortening makes it appear longer along its north-south axis. The apparent shape of Mare Crisium and its nearness to the eastern edge of the Moon is a good naked eye guide to the extent of lunar libration.

Libration, the apparent slight rocking of the Moon from side to side during the month, allows a total of 59 per cent of the Moon's surface to be seen over time, while the remaining 41 per cent of the Moon (the true far side) is always unobservable. Libration affects the way in which we view lunar features, especially those lying nearer to the Moon's edge, and whenever Mare Crisium is on show, those with a keen naked eye will be able to determine the state of the Moon's libration.

Libration favouring the eastern edge of the Moon will show Mare Crisium to good effect, while a strong libration favouring the Moon's western edge will push Mare Crisium near the Moon's edge, foreshortening it to a considerable extent. As the Moon waxes (grows in size) in the evening skies, libration greatly favours Mare Crisium, while the remainder of the lunar cycle sees Mare Crisium gradually being pushed ever-closer to the Moon's edge.

The neat, oval-shaped Mare Crisium produces a noticeable 'dent' on the morning terminator of the young lunar crescent, visible without optical aid to those with keen eyes. Fully exposed to the morning sunshine a day later, Mare Crisium makes a spectacular sight through any optical instrument, looking like a large flooded crater (which is exactly what it is). Mare Crisium has lofty western mountain borders whose clean-cut cliffs shine brilliantly in the morning light.

Under a low illumination, a concentric system of wrinkle ridges comes into view. These ridges average about 50 kilometres (31 miles) from the mare border, forming a disjointed internal ring. Dorsum Oppel, the most prominent of these wrinkles, links with the flooded crater Yerkes (36 kilometres, or 22 miles) in the west and curves around the northwestern periphery of the mare for 300 kilometres (186 miles), where it is intercepted by half a dozen narrow wrinkles that cross the mare from the northwestern border.

In the northeast lies the narrower Dorsa Tetyaev (150 kilometres, or 93 miles, long) and Dorsa Harker (200 kilometres, or 124 miles). As the Sun climbs higher, numerous light-coloured rays cross Mare Crisium's mottled surface, notably those from the bright impact crater Proclus (28 kilometres, or 17 miles, across) just beyond its western border. Several impact craters dot its surface — Picard, Peirce and Greaves can be seen under a midday illumination.

A couple of days after full Moon, Mare Crisium's western border casts shadows as its eastern reaches begin to darken with the nearing sunset terminator, the mountains of its eastern border glinting in the setting Sun. A considerable breach exists in the eastern mountain border where the mare lavas have flowed into outlying craters and valleys, notably Mare Anguis (Serpent Sea), one of the smallest lunar maria which is an irregular, dark patch measuring about 200 kilometres (124 miles) from north to south.

A large mountainous headland, Promontorium Agarum, projects into Mare Crisium from its southeastern shore. There's far more to glimpse in and around Mare Crisium throughout its two weeks in the Sun during each lunation — it truly is a fascinating feature to study.

Top tip!

Mare Crisium will make a spectacular sight through any optical instrument, as it is fully exposed to the morning sunshine and will appear as a large, flooded crater. A Moon filter will improve contrast, toning down any glare that often washes out intricate features.

Mare Humboldtianum

A favourable libration makes this elusive lunar sea an intriguing target

Now it's time to turn to the very edge of the Moon and take a look at one of the most elusive lunar seas visible from Earth — Mare Humboldtianum. Located on the Moon's northeastern limb, Mare Humboldtianum is a dark patch of lava some 270 kilometres (170 miles) across on the lunar nearside, whose eastern edge just touches the 90 degrees east line of longitude.

Since the Moon rotates once on its axis in precisely the same time as it takes to revolve around Earth (keeping the same face turned towards us), it might be thought that Mare Humboldtianum should always be on view whenever it is illuminated by the Sun. But this isn't the case, owing to a phenomenon known as libration.

Libration produces an apparent slow rocking motion of the Moon, a phenomenon that allows a total of 59 per cent of the Moon's surface to be seen over time, while the remaining 41 per cent of the Moon — the true far side — is perpetually hidden from our gaze. Libration has a number of causes, but the main effect is caused by the Moon's elliptical orbit around Earth combined with the steady rotation of the Moon on its axis each lunar month. Libration can bring features on the mean far side into our telescopic sights, and it can also work the other way, pushing features that are near the mean lunar limb out of sight — the latter applies to Mare Humboldtianum.

Under an unfavourable libration (where the Moon's southwestern limb is well-seen), Mare Humboldtianum is shunted onto and beyond the northeastern limb, rendering it virtually unobservable. However, a favourable libration (combined with a favourable illumination) brings the Moon's northeastern limb regions into view, and Mare Humboldtianum is pretty easy to spot, even through binoculars.

The latter circumstance takes place in early March, so it's a great opportunity to take a look at one of the Moon's smaller and lesser-known seas. First, a little historical perspective. Mare Humboldtianum is a dark patch of lava (270 kilometres or 170 miles across) that fills the central regions of a much larger ancient impact basin (around 650 kilometres or 404 miles across) whose eastern reaches extend well onto the Moon's true farside. The basin-forming impact took place around 3.8 billion years ago.

Later impacts have scarred Humboldtianum, with light coloured ejecta from local impact craters and the huge 200-kilometre- (124-mile-) wide crater, Belkovich that intrudes upon Mare Humboldtianum's northeast flank, straddling the mean lunar near and far sides.

The feature was named Mare Humboldtianum in 1837 by German astronomer and selenographer Johann Mädler in honour of his compatriot, the naturalist and explorer Alexander von Humboldt. Obviously, Humboldt's explorations of unfamiliar terrestrial continents in the late 18th and early 19th centuries formed a symbolic analogy to Mädler's own lunar surveys, and therefore Mare Humboldtianum represented a physical link between the known and (then) unknown hemispheres of the Moon. Viewed from above, Mare Humboldtianum appears as a broad crescent, and it was first pictured from space by the Soviet probe Luna 3 back in October 1959.

A favourable libration of the Moon's northeastern limb between the Moon's age from 3 to 13 days in its waxing phases enables Mare Humboldtianum to be seen very well.

The best evenings to spot the Mare Humboldtianum is when libration is at its maximum for the feature. Binoculars will easily realise it as a dark patch near the northeastern (upper-right) edge of the Moon, and a telescopic view will show considerable detail, although it is illuminated by a relatively high Sun and no shadows will be cast by any of its relief features.

Top tip!

You should observe this lunar sea during the libration of the Moon's northeastern limb in the Moon's waxing phases. A Moon filter will improve contrast, toning down any glare that often washes out intricate features.

© NASA

Mare Undarum

A favourable libration provides an ideal opportunity to view the lava-filled craters on the Moon's eastern edge

Earlier on, we took a look at Mare Crisium (the Sea of Crises), a vast oval-shaped dark plain near the Moon's northeastern edge — a lunar sea (or maria in Latin) so sizeable that it can easily be seen with the keen, unaided eye as a dusky spot near the lunar limb. Mare Crisium may be looked at as a very large impact crater, whose central plains have been covered by lava that has welled up from beneath the lunar crust at some point in time after impact.

A phenomenon known as libration — an apparent slight rocking of the Moon from side to side during the month — enables Mare Crisium to appear more prominent from time to time. However, Mare Crisium lies entirely on the Moon's nearside and its oval shape can always be observed when the area is illuminated. But there also exists a cluster of smaller lunar seas also visible, which are located to the east of Mare Crisium. None of these seas take the classic oval form of Mare Crisium, for each is somewhat patchy and irregular in outline.

Located around 100 kilometres (62 miles) southeast of Mare Crisium lies Mare Undarum (Sea of Waves), which is a collection of lava-filled craters. Like Mare Crisium, Mare Undarum is positioned firmly on the Moon's nearside and it therefore always remains visible whenever it is illuminated, despite the effects of libration. Mare Undarum's outline is rather irregular, roughly 100 kilometres (62 miles) from north to south and 200 kilometres (124 miles) from west to east. Even at the extreme of libration, Mare Undarum can be seen in its entirety — although foreshortened to some degree — near the Moon's edge. However, two eastern seas lie much further east than Mare Crisium on the lunar disc, both of which are sizeable lunar marias that actually straddle the line of 90 degrees east — which separates the mean nearside and farside of the Moon. These features are Mare Marginis (the Border Sea) and Mare Smythii (Smyth's Sea).

Mare Marginis lies due east of Mare Crisium. Irregular in outline and measuring 360 kilometres (224 miles) from east to west, it occupies the libration zone. Extreme librations see the mare disappear around the limb nearly completely, but it can usually be glimpsed as a narrow elongated dark area on the eastern limb during the first half of the lunation up to Full Moon. Located on the eastern limb, Mare Smythii is around 200 kilometres (124 miles) across and viewed from above it assumes a roughly circular (though poorly-defined) outline.

Although the whole of Mare Smythii is presented towards the Earth during a favourable libration, its near limb position makes it appear very foreshortened. Like Mare Marginis, the sea comprises of many large, lava-filled craters and it can disappear beyond the eastern limb during an unfavourable libration.

"Extreme librations see the mare disappear around the limb nearly completely"

Mare Imbrium

Explore the Moon's Sea of Showers

One of the most iconic images in the history of space exploration dates back to 1902. In his 1902 film Trip to the Moon, one of the first sci-fi films ever made, pioneer film-maker Georges Méliès fired a bullet-shaped capsule to the Moon from a giant cannon, and that iconic image shows how it landed in the poor Man in the Moon's right eye. Today when we look up and see the Man in the Moon staring back at us, his right eye thankfully seems to have healed — and we now know it as Mare Imbrium, the Sea of Showers.

Like all the major lunar seas, Imbrium is an impact basin — a wide, lava-filled wound caused by the impact of an enormous body many billions of years ago. Lunar scientists estimate that Imbrium was blasted out of the lunar crust around 3.9 billion years ago, by a protoplanet 250 kilometres (155 miles) wide. One cataclysmic impact excavated a hole 1,225 kilometres (760 miles) wide, which we call Mare Imbrium today.

Although it looks impressive from Earth through a telescope, studies from the many crewed and robotic spacecraft which have flown over it have revealed Mare Imbrium is a truly fascinating place. A vast mottled plain of ancient frozen lava, dotted with huge craters, Imbrium is bordered by the jagged Apennine Mountains to the east, the dark-floored crater Plato to the north and horned crescent basin of Sinus Iridium catch the observer's eye as they gaze at Imbrium through even a small telescope. When sunlight strikes Mare Imbrium at a shallow angle, it reveals that its seemingly-flat floor is wrinkled and rippled like an un-ironed sheet.

Because it offers so much fascinating and diverse geology, over the years several major space missions have landed within Mare Imbrium's borders. In 1970, the Russian Luna 17 probe landed the first Lunokhod rover on the Moon inside Mare Imbrium. The following year the crew of Apollo 15 set down next to the meandering 1.6-kilometre- (one-mile-) wide channel Hadley Rille. Here Dave Scott and James Irwin spent almost three days studying the Moon's geology and collecting rock samples, including the bright 'Genesis Rock'. They also enjoyed long, bumpy drives across the lunar surface in the first lunar rover to go to the Moon. More than 40 years later, wheels left tracks across Mare Imbrium again when the Chinese Chang'e 3 lander set down on the sea's dusty surface and the Yutu rover trundled down its ramp and out onto the boulder-strewn surface, sending back hundreds of beautiful images.

Mare Imbrium remains in darkness until the terminator first touches the peaks of the Appennine Mountains on the impact plain's eastern rim. A few days later, Imbrium is half-illuminated and a couple of days after that, it is all bathed in sunlight and its features will stand out.

This will be the best time to see the trio of prominent craters on its eastern side — Archimedes and its smaller more northerly neighbours Autolycus and Aristillus — and the ridges and wrinkles on its dark floor: Imbrium then remains in the light for ten days until the sweeping of the terminator across the face of the Moon begins to surrender it to the long lunar night and the Appennine Mountains go dark once more.

Top tip!

Observe Mare Imbrium as soon as the terminator reaches it for the best views of the subtle features like wrinkles and small craters.

© NASA

Mare Orientale

Catch a fleeting glimpse of one of the Moon's most incredible features

Top tip!

If you have a Moon filter, use it when looking for Mare Orientale. The contrast will make the feature much easier to see.

Most of the lunar features we visit on our tours are pretty obvious and easy to see — huge pit-like craters, long chains of jagged mountains, vast dark seas, etc. This target is much more challenging and much harder to see. In fact, you'll only be able to glimpse it for a couple of days. Why? Because it is a feature usually invisible from Earth.

Mare Orientale and its surrounding rings of mountains form one of the largest, most dramatic features on the Moon. If it was on the Earth-facing side of our celestial companion it would dominate its face, and would quite possibly have affected the development of many of our cultures and religions. Unfortunately, it was blasted out of the Moon just far enough around the western limb that it is usually hidden from our view. However, occasionally the libration — or wobbling — of the Moon allows us to sneakily peek a short distance around the western limb, and offers lunar observers a tantalising, fleeting glimpse of this fascinating feature.

Only the wide-with-wonder eyes of Apollo astronauts and the clicking cameras of robotic spacecraft have seen, and photographed, Mare Orientale in all its glory. They gazed down at one of the youngest impact features on the Moon, blasted out it around 3.8 billion years ago by an asteroid more than 60 kilometres (37 miles) wide.

That brutal collision painted an enormous bullseye target on the Moon — a 327-kilometre- (203-mile) wide dark sea, or mare, of frozen lava, surrounded by three concentric rings of crater-pocked mountains that make the whole feature more than 900 kilometres (559 miles) across. Just imagine if Mare Orientale had been formed on the side of the Moon facing Earth — our natural satellite would resemble a huge eye, staring down at us from the sky. Now imagine seeing that painted red, bloodshot during a total lunar eclipse. It's fascinating to wonder how much fear would that have caused, and how such an ominous sight would have affected our species' cultures and religions, isn't it?

Sadly, our observing windows for this fascinating feature are few and far between — and brief, too — so any opportunity to catch even a fleeting glimpse of Mare Orientale should be grasped. When the Moon's libration woozily swings the western side of the Moon towards us and Mare Orientale, its surrounding mountain rings will become visible — but only through a good pair of binoculars or preferably a telescope, and even then only as an area broken up into light and dark lines close to the lunar limb. Even so, just seeing the enigmatic Eastern Sea at all is a thrill, so cross your fingers for clear skies at that time.

Where exactly should you look? The easiest way to find Mare Orientale is to go back to basics and imagine the Moon's face as a clock face. Mare Orientale will tilt towards us, so when libration is at its most pronounced, aim your binoculars or telescope towards the 8 o'clock position where you will see the dark-floored crater Grimaldi. Beneath and to the left of Grimaldi, right on the limb, you will see what looks like a number of dark lines, almost like scratches on the Moon — these are Mare Orientale and its numerous mountains.

Of course, the more magnification you use with your telescope, the more detail and structure you will see — but no matter how high you go, you won't be able to see the circular shape of the feature. However, in a way that doesn't really matter. What does matter is that you will be seeing something that is usually hidden from our view. You'll be able to feast your eyes on something that most sky-watchers, and many Moon observers, have never seen.

Sea of Tranquility

Explore the dark sea that has accommodated previous Apollo missions

The Sea of Tranquility, also known as in Latin as Mare Tranquillitatis, is one of the best known lunar features around due to it being easy recognisable and it being the landing site for the historic Apollo 11 landing. This mare — the name given to large, dark, basaltic plains — was formed from the ancient eruptions of lunar volcanoes roughly 4 billion years ago, before the Moon settled into the ball of rock seen today.

The Sea of Tranquility stretches approximately 870 kilometres (540 miles) in diameter, which is roughly the same distance between San Francisco and San Diego in California, United States. It is not hard to spot this mare when the area is illuminated, and you can find it using a modest pair of binoculars.

It would be best to observe this region when the Moon is at a high level of illumination as the surrounding white-greyish land will emphasise the dark basaltic area. In terms of location, it can be found along with another connecting mare, Mare Serenitatis or Sea of Serenity, flaunting a snowman figure with an opposing colour scheme to it. Other notable features around it include the much smaller Mare Nectaris or Sea of Nectar. As this area is much darker, its harder to decipher any craters or distinguishable landmarks, but its contrast in colour makes for an enjoyable view, particularly around the rim of the seas.

It is thought that this region was formed during the 'pre-Nectarian' period, which is a timescale spanning between 4.5 and 3.9 billion years ago. Fast-fast-forwarding to present day, this land is peppered with tiny craters from impacts over the years as well as ridges, grooves and volcanic channels from previous geological activity. However the stepping on the surface has been described as like walking on powder, as the face of the Moon is covered in very, very fine-grained rock.

This site is probably most notable as the area where Neil Armstrong and Buzz Aldrin touched down on 20 July 1969 aboard Apollo 11's Lunar Module Eagle. When they stepped off Eagle, they became the first humans to step foot on a different celestial body, and this all occurred at Base Tranquility, as they named it. After approximately six hours collecting samples and taking photographs, they left behind an American flag, as well as a patch honouring the fallen heroes of Apollo 1 and the rest of the Lunar Module, with a plaque on the leg of Eagle that read: "Here men from the planet Earth first set foot upon the Moon. July 1969 A.D. We came in peace for all mankind."

Spotting the landing site is an enjoyable task, as is finding the other five Apollo landing sites, and this can be done with the use of a decent telescope capable of resolving craters of the surface. As for finding the spot where Apollo 11 touched down, it would be easier to find the nearby crater Theophilus first. After this crater is located, follow an imaginary line up towards the Sea of Tranquility where the smaller, yet still distinct, crater Moltke lies. Just northwest of Moltke is the landing site, and if the telescope and user's eyesight is good enough, the three incredibly small craters — Aldrin, Armstrong and Collins, the latter named after Michael Collins who was commanding the Command Module orbiting the Moon on the mission — can be found.

Before this courageous feat of exploration, Mare Tranquilliutatis has been the subject of a couple of investigations. For instance, in 1965, NASA's Ranger 8 spacecraft obtained the first close-up images of the Moon and finished its mission by crashing into the Sea of Tranquility.

Another NASA lunar exploration mission visited the same basaltic sea, but the Surveyor 5 made a much more controlled landing in 1967. Due to the fact that this region has become one of the most explored areas of the Moon, it has managed to capture the imagination of people worldwide, and its presence has been used in many different works of art from songs to books.

© NASA

"Spotting the landing site is an enjoyable task, as is finding the other five Apollo landing sites, and this can be done with the use of a decent telescope"

Top tip!

Finding the Apollo 11 landing site is easier when the Moon is finely illuminated in order to distinguish the small number of nearby craters.

Mare Serenitatis

When the Moon is full, enjoy views of one of the lunar surface's ancient lava seas

Look at the full Moon with your naked eye or a pair of binoculars on a frosty winter's night and you will see areas of light and dark. The light areas are known as the 'lunar highlands' and they are the oldest landscapes on the Moon, regions of mountainous, rugged terrain. The dark areas are the lunar 'seas', much younger than the ancient highlands. However, these seas are not like the huge stretches of water found on Earth. Instead, they are enormous plains of ancient, frozen lava.

At full Moon you will be able to see this Moon Tour target, Mare Serenitatis — the Sea of Serenity. Mare Serenitatis was formed around 3.8 to 3.9 billion years ago, when the Moon was experiencing a period of intense and heavy bombardment. As the Moon, still young itself, was being pummelled by countless planetesimals, asteroids and chunks of space debris left tumbling around the infant Sun after the birth of our Solar System, an enormous, Moon-shuddering strike blasted a huge impact basin out of the lunar surface. This basin then filled with lava that sloshed over its floor, covering it and any features on it.

Today we know that vast lava plain as Mare Serenitatis, one of the most obvious features on the Moon's surface. We also know that Serenitatis is approximately 674 kilometres (419 miles) across at its widest point, and more than 2 kilometres (6,500 feet) deep at its deepest point. Although the sea itself is very flat and smooth, its surroundings are not. Its western edge is marked by the jagged crater-pocked Appenninus and Caucusus Mountains, and more tortured, broken terrain lies on its southern border.

Although no crewed or uncrewed spacecraft have landed on the vast plain itself, in 1972 the final Apollo mission, Apollo 17, touched down on its south-eastern border, near the Taurus-Littrow Valley. The lunar module Challenger carried astronauts Eugene Cernan and Harrison H. Schmitt down to the surface, where they drove a lunar rover for almost 40 kilometres (25 miles), visiting different sites and collecting a wide variety of rocks. A year later the Russian Luna 21 robot probe, carrying the Lunokhod 2 rover, touched down in the same area.

Looking at Mare Serenitatis through a small telescope, you immediately notice two features upon it. First, a small, round crater just below its centre — this is 16-kilometre- (ten-mile-) wide Bessel. Looking at it, you'll notice the second feature; the crater sits in the centre of a bright 'ray' that looks like a white chalk line, pointing down towards the 27-kilometre- (18-mile-) wide crater Menelaus on the south 'shore' of the sea. This is the 'Bessel Ray', thought to be a splash of material from the impact that blasted the crater Tycho thousands of kilometres away to the south.

Scanning Serenitatis through a telescope's high-power eyepiece will reveal it is marked with many wrinkles, ridges and crags, including a long, meandering ridge which runs from north to south on the sea's eastern side. There are also around a dozen other craters scattered across it, but they are all small and of no real significance.

When the Moon is almost full the sea will be in view until the terminator — the line between night and day — begins to slide across it. As the darkness encroaches, it will only be half-visible, before being completely swallowed up by the relentlessly increasing shadow. It will remain hidden from view until the terminator's advance across the Moon allows sunlight to bathe the sea's shore once again. Then, when the waxing Moon hangs low in the western sky after sunset, the whole of Serenitatis will be in full view once more.

Top tip!

If you look at Mare Serenitatis when it is close to the terminator, you will see more detail and surface relief to it.

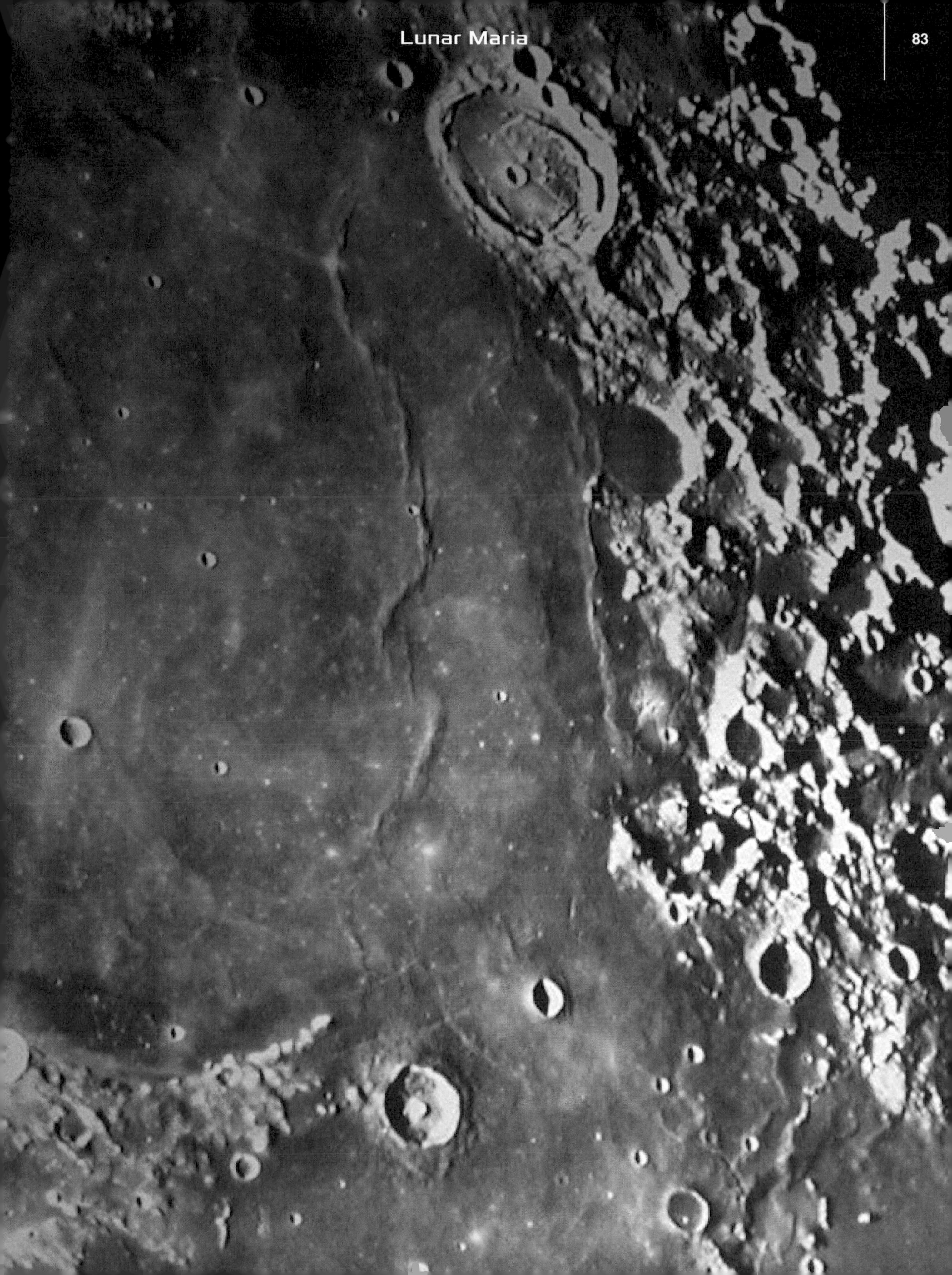

Other Landmarks

Uncover some of the other features of the Moon's surface with this handy guide

The Apennine Mountains

Take a flight over one of the Moon's most spectacular mountain ranges

If you've ever seen the famous 'man in The Moon' when looking up at a bright, silvery full Moon on a frosty winter's night, you'll be able to find this lunar feature without any difficulty. The two dark eyes — the ancient frozen lava seas of Mare Imbrium and Mare Serenitatis — are very distinctive, with a bright nose curving down between them. This nose is actually one of the most impressive mountain ranges on the Moon: the Apennine Mountains.

Named after the Apennine Mountains in Italy, this mountain range is so close to the centre of the side of the Moon facing Earth that it is best seen when the Moon is at or close to first or last quarter, when it lies close to the terminator. With sunlight hitting it at an angle, the mountains are stunning through even a small telescope — a curved, jagged line of hills and peaks looks like the fossilised spine of a dinosaur sticking out of the crust. In comparison, at full Moon it is reduced to little more than a bright grey-white line.

The Apennines are magnificent. Stretching from the crater Eratosthenes in the south, up to the craters Archimedes, Aristillus and its near neighbour Autolycus further north, the mountain range is a curving chain of jagged rock that stretches for more than 600 kilometres (373 miles) across the Moon, further than the distance from London to Dundee in Scotland. The tallest Apennine peak, Mons Huygens, is also the tallest mountain on the Moon. With a 5.5-kilometre (3.4-mile) high summit, Mons Huygens is higher than Mont Blanc in the Alps.

Geologists believe that the Apennines were created 3.9 billion years ago during the colossal asteroid impact that formed Mare Imbrium, the dark lava sea that lies to their west. When looking at the Appennines through a telescope, it is thrilling to imagine their formation all that time ago; watching the huge chunk of space rock slamming into the Moon, blasting out the dark Imbrium basin, flooding it with glowing lava and pushing up the crust into the mountains we see today.

Through a telescope eyepiece at medium-to-high magnification, the mountain range is broken up into many hills, hummocks and peaks. There are several gaps along the chain, perhaps the most prominent being a steep-walled valley to the north of Mons Wolf. At the northern end of the range, in the shadow of towering Mons Hadley, a narrow canyon, or rille, snakes its way across the lunar surface.

Through a telescope at high magnification, it looks like no more than a black hair lying on the grey ground, but this is the famous Hadley Rille, which served as the landing site of the Apollo 15 mission. In July 1971, astronauts Dave Scott and James Irwin explored this area for almost three days. During their long stay on the Moon, they used the famous lunar rover for the first time, driving 28 kilometres (17 miles) across the Moon's surface, carrying out a detailed geological survey and gathering precious rock samples. In fact, it was here that Dave Scott found the most famous and most scientifically important rock collected during the whole Apollo programme: the 'Genesis Rock' — a lump of crystalline rock that is more than 4.5 billion years old, formed only 100 million years or so after the birth of the Solar System. So why not have a go at finding it for yourself?

Top tip!

Start looking for the Apennines after sunset on 6 December. How soon will you see their tallest peaks catching the rays of the rising Sun?

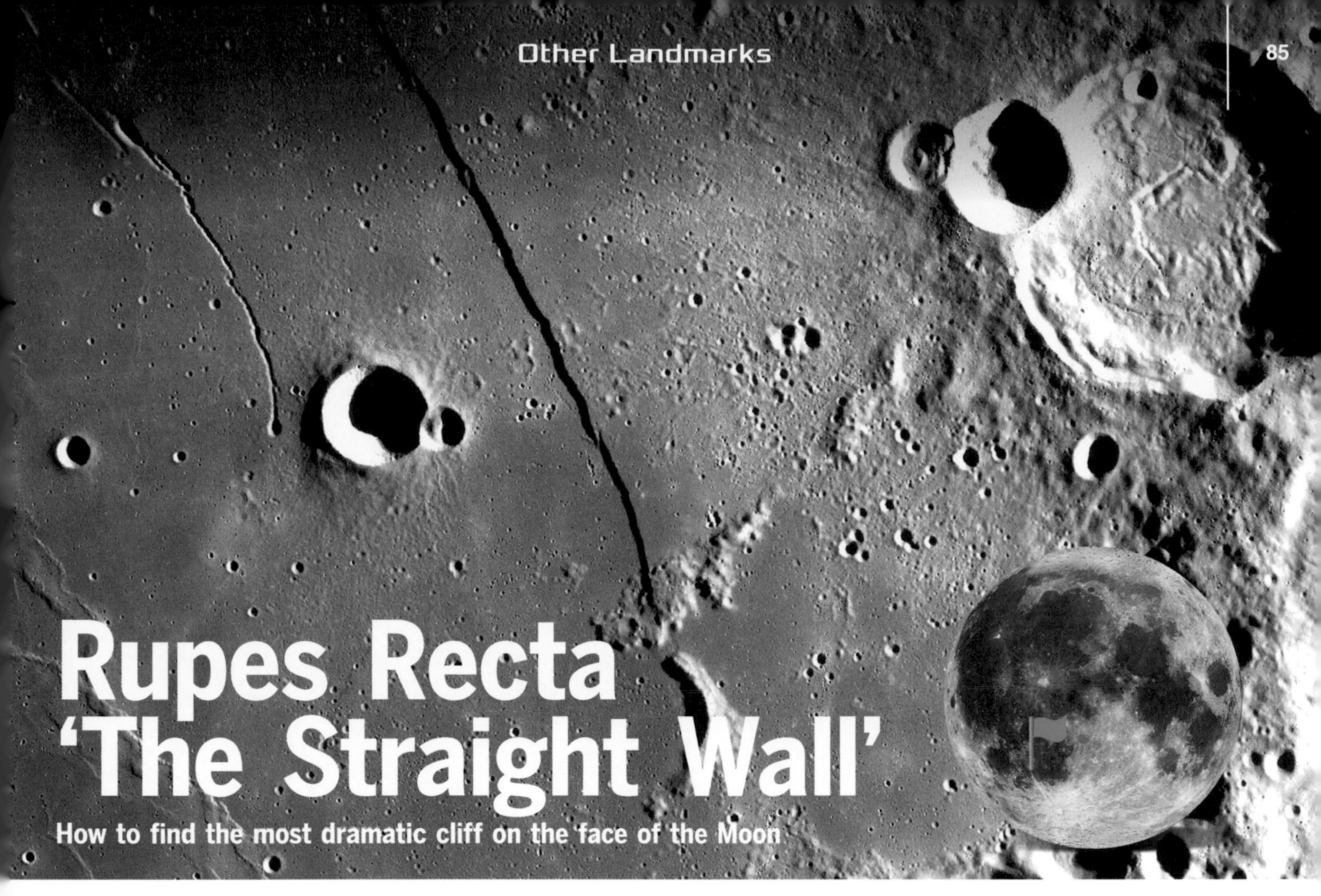

Rupes Recta 'The Straight Wall'

How to find the most dramatic cliff on the face of the Moon

If you have a small telescope or a powerful pair of binoculars and look to the lower left of the great triple crater chain of Ptolemaeus, Alphonsus and Arzachel, you will see (depending on the time of the month) either a short, dark line or a short, bright line. Moon atlases and phone apps identify it as 'Rupes Recta'. While it doesn't look much at first glance — nothing more than a dark pencil line or a white chalk scratch drawn on the Moon's ash grey face — Rupes Recta has another name, and is one of the most famous and beloved features on the whole of the Moon: 'The Straight Wall'.

Of course, it's not actually a wall; it wasn't built by tea-gulping lunar labourers leaning on shovels! Rupes Recta was built by the forces of nature, it is an enormous scarp, a region where part of the Moon's surface dropped dramatically away, forming a steep cliff. The cliff itself is very narrow, barely a couple of kilometres wide and nowhere near as wide as the terraced rims of those three giant craters blasted out of the Moon to its north. It's not all that tall either: with a maximum height of around 450 metres (0.28 miles), it's roughly as tall as the London Eye, or two nuclear submarines that are balanced end to end.

Although the Straight Wall gives the impression of being a towering cliff face, it's not. Pre-Apollo space artists depicted the Straight Wall as vertical, a frozen tsunami wave of grey lunar stone, but if you stood at the base you would see the slope rising at an angle of only 30 degrees or so. Rupes Recta doesn't turn out to be very wide, or very high; so what's all the fuss about?

Well, the Straight Wall's remaining claim to fame is that it's very long. Stretching more than 110 kilometres (68.3 miles) across the lunar surface, it would reach from London to the Isle of Wight if placed on the Earth, or from Carlisle to Edinburgh if you prefer a more northerly comparison. That's so long that it would have taken Apollo astronauts more than eight hours to trundle and bounce from one end to the other in the lunar rover, which was used on the Moon in the later missions.

The Straight Wall becomes visible when the Moon is just past its First Quarter phase and low in the southwestern sky after sunset, to the upper left of Saturn. With sunlight illuminating it from the east, the cliff face will be in shadow and appear as an obvious short, dark line to the lower left of Arzachel, very close to a small, deep pit of a crater called Birt. As sunlight creeps across the Moon's face and the cliff's shadow retreats, the Straight Wall will slowly turn into a bright line which seems to sink down into the Moon, becoming harder and harder to see until it's barely visible when our natural satellite is full.

The best time to see the Straight Wall is when the Moon has just passed Last Quarter and is shining between the famous Pleiades and Hyades star clusters of Taurus. Now illuminated from the west, Rupes Recta's cliff face will be bathed in full sunshine, making it appear as a strikingly bright line etched into the darker surface of the Moon — looking as if it has been dug out of the ancient frozen lava flows of Mare Nubium by the tip of a knife's blade. At this time it will be visible with a good pair of binoculars, but telescopes will offer the best views, the higher the magnification the better. All too soon Rupes Recta will be lost from view, as local sunset plunges it into darkness, hiding it from our gaze for around a fortnight

One day in the future, astronauts will surely come to the top of the Straight Wall and gaze down on the Moon's magnificent desolation from its lofty heights. Until then, we can enjoy gorgeous views of it from Earth, with the most modest observing equipment.

"It would reach from London to the Isle of Wight"

Top tip!

The Straight Wall is easiest to see when it is near the terminator (the line between night and day).

© NASA

Top tip!

As with other lunar features, you'll enjoy your best view of Mons Pico when it is close to the terminator.

Mons Pico

Explore one of the most isolated mountains on the lunar surface

This destination is a small, unimposing mountain towards the north of the Moon. Mons Pico — or Mount Pico — is a small nub of a mountain that is 25 kilometres (15.5 miles) long, 15 kilometres (9.3 miles) wide, and 2.4 kilometres (1.5 miles) in height. However, while this might sound impressive, Pico isn't as high as even the 100th-tallest mountain on Earth. In fact, it is only around the same height as Mount St Helens, the American volcano that famously blew its top in May 1980.

So, clearly it's not Pico's height that makes it worth seeking out. What then? It's the mountain's isolation. Mons Pico can be found in the far north of the huge Mare Imbrium basin, a short distance to the south of the famous dark-floored crater Plato. It stands out — and is so easy to find — because there really isn't much else around it. To the northwest is a small v-shaped range of much smaller peaks, the Tenerife Mountains, and directly north of Pico is a short chain of craters that look like holes left in an antique table by woodworm, but they're so tiny that you'll need very high magnification to see them.

Pico might not be as high as other lunar mountains, but just like its taller relations it casts a shadow, and when the terminator — the line between night and day — is nearby that shadow can be very striking, and makes Pico stand out very clearly from the dark, rolling plain of Mare Imbrium. At full Moon, with the baking Sun straight overhead and all surface relief banished by the harsh glare, the mountain is reduced to a bright spot, as white as a dot of correction fluid on paper. Conversely, when sunlight strikes Mons Pico at a shallow angle, the mountain casts a shark-tooth of shadow that makes it stand out starkly, even at low magnification through a small telescope.

Pico might not be well known to casual observers of the Moon, but it is to science-fiction fans and space-exploration enthusiasts. Iconic space artist Chesley Bonestell painted the mountain several times in the giddy pre-Apollo years; it features in the ground-breaking 1951 book The Conquest of Space and also in one of his illustrations for the 'A Trip to the Moon' feature in the 1946 issue of Life magazine. And fans of Arthur C Clarke's Odyssey series will recognise Mons Pico as the site of humanity's repository of biological and computer viruses.

So when can you see Pico for yourself? The best time is usually when the terminator has just swept over the mountain and it will be a striking feature, looking like a sharp piece of bone sticking out of the Moon with a dark shadow cast behind it to the west. For the next few days after that, it will be bathed in more and more sunlight. The shadow will gradually shrink back to the foot of the mountain until the full Moon, when Pico will look like a white dot beneath the dark blue-grey circle of Plato.

Pico will begin to become more apparent again as the terminator creeps back over it, this time from the other direction. By now the Moon will be at its waning gibbous phase in the pre-dawn sky, so seeing Mons Pico emerging from the darkness will mean having to get up early, but as you look at it you'll be able to see Mars, now very bright, in the same area of the sky too.

"It will be a striking feature, looking like a sharp piece of bone"

Montes Alpes

This lunar edition of the European mountain range makes for an entertaining observation challenge

Montes Alpes, also referred to as the lunar Alpes after the famous Alps mountain range in Europe, is a selection of mountains that can be found along the northern part of the Moon's near-side. More specifically, this range can be found in between the Plato and Cassini craters and along the rim of Mare Imbrium, which is Latin for Sea of Showers or Rains.

This range is best viewed in the first nine days after a New Moon when the moon is in its waxing gibbous phase, where a large majority of near-side face is about 73 to 86 per cent illuminated. However, the terminator definitely provides the best conditions to observe the tremendous sights of Montes Alpes. The inner walls of the mountains are steep and well-defined, as opposed to the outer walls that appear to be more broken as the elevation decreases away from Mare Imbrium.

The mountain range spans across roughly 330 kilometres (200 miles) of the Moon's surface, but they weren't formed in the same way as the mountains ranges on Earth. On Earth, churning magma at the centre of Earth moves crustal plates, with mountains forming from the collisions of these. Their lunar counterparts, however, were formed as a consequence of enormous collisions by large asteroids that were abundant in the younger Solar System. When an enormous asteroid hit the Moon, the heat produced from the impact was able to melt and mould the surface, creating the famous mares and mountains along their rims.

This is no different for Montes Alpes, which is believed to have formed roughly 3.85 billion years ago, but this was the result of a collision with a protoplanet 250 kilometres (155 miles) wide. This impact created a saucer-shaped basin with many mountain ranges formed around the rim, including Montes Jura, Montes Apenninus, Montes Carpatus, Montes Caucasus and, of course, Montes Alpes.

The mountains of the lunar Alpes range in height from 1,800 to 2,400 metres (5,900 to 7,900 feet), paling in comparison to some of the mountains that constitute the Earth-based Alps as about 100 peaks are higher than 4,000 metres (13,000 feet).

Separating the mountain range is the wide-rift valley known as Vallis Alpes, which can be located in the northwestern third of the range. This 170-kilometre (100-mile)-long valley provides a ten-kilometre (six-mile) division in Montes Alpes and was formed from the flowing of ancient lava that creates this flat valley floor. In fact astronomers believe this valley has magma flowing between Mare Imbrium and Mare Frigoris, but it is also possible that stress fractures could have formed Vallis Alpes. Running through the centre of Vallis Alpes is a narrow, long depression called a rille, but unfortunately it is far too small to be seen with any ground-based telescope. Much like the Alps on Earth, the mountain that reins supreme across the lunar Alpes is Mont Blanc, that has the highest peak of around 3,500 metres (11,500 feet) and is about 25 kilometres (15 miles) wide.

This is 1,300 metres (4,200 feet) smaller than the Mount Blanc on Earth, and doesn't make for a good winter holiday. As there is no atmosphere, no snow and the gravity is incredibly weaker than Earth's, a 70 kilogram (150 pound) skier would only weight 10 kilograms (25 pounds) as they ski down these mountains. It would be best to visually explore the lunar Mont Blanc and the rest of Montes Alpes with a telescope that provides a magnification of 100x. If you wanted to up the ante and zoom in either further, a 200x magnification can unveil finer details of the mountain range.

Top tip!

Observe Mare Imbrium as early in June as you can for the best views of the subtle features on its floor, such as wrinkles and small craters.

© NASA

Moon Landings

106

"The Apollo missions were a spectacular success and represented a golden age"

92

138

98

Project Apollo

NASA's groundbreaking effort to reach the Moon

1961-1966

SA and AS unmanned missions
In preparation for the manned Moon missions, NASA conducted a series of tests using various iterations of the Saturn rocket, in order to practice launch, Low Earth Orbit, re-entry and mission aborts.

27 January 1967

Apollo 1
The first manned mission.

9 November 1967

Apollo 4
After Apollo 1's launch test disaster, the nomenclature of subsequent missions was changed. No Apollo 2 or 3 existed; the next mission in the programme, a Type-A unmanned flight, was the first test of the Saturn V rocket that would eventually take man to the Moon.

22 January 1968

Apollo 5
The next unmanned mission was a Type-B on a Saturn IB rocket and marked the first flight of the Lunar Module, including successful tests of its ascent and descent engines, and a simulation of a landing abort, referred to as a 'fire-in-the-hole' test by NASA engineers.

4 April 1968

Apollo 6
The final unmanned Apollo mission was used to test the Saturn V's ability to propel the spacecraft into trans-lunar injection (TLI). The flight experienced problems from the start, including a vibration problem called pogo oscillation that damaged the fuel lines. It was also the only Saturn V-launched Apollo flight with a white Service Module — all the others were silver in colour.

11 October 1968

Apollo 7
On a warm October day, the first manned Apollo mission launched from Cape Canaveral and it was a great success.

21 December 1968

Apollo 8
The astronauts in this mission were the first to see the far side of the Moon.

3 March 1969

Apollo 9
This mission spent a total of ten days in orbit.

18 May 1969

Apollo 10
See page 96.

16 July 1969

Apollo 11
See page 98.

14 November 1969

Apollo 12

Apollo 12 was a stormy flight. It launched amid thunder and was struck by lightning, which overloaded the telemetry systems and caused them to fail. Flight controller John Aaron had seen this error happen before in a simulation, leading him to give the unusual command "Apollo 12, Houston. Try SCE to Auxiliary. Over", which confused two of the three astronauts aboard. Lunar Module pilot Alan Bean, however, remembered a similar simulation from his own training and knew where the obscure switch was located. He flipped the systems to a backup power supply as per Aaron's request, and the telemetry data came back online. With the problem solved, the flight went on to make the first precision lunar landing in — where else? — the Ocean of Storms.

11 April 1970

Apollo 13

See page 106

31 January 1971

Apollo 14

See page 108

26 July 1971

Apollo 15

See page 110

16 April 1972

Apollo 16

The first mission to land in the lunar highlands, Apollo 16's flight was largely textbook, despite a launch delay due to technical problems. The crew conducted three moonwalks on the lunar surface, totalling more than 20 hours and exploring both on foot and in the lunar rover, and conducted experiments on the Moon and in space on their return journey home, including deploying a small subsatellite that was intended to orbit the Moon but crashed into its surface just 35 days later, inadvertently leading NASA to the realisation that there are only four so-called 'frozen' lunar orbits in which an object can orbit the Moon indefinitely, at 27, 50, 76 and 86 degrees inclination: gravitational anomalies called mascons drag objects in other low lunar orbits inexorably towards the Moon's surface.

7 December 1972

Apollo 17

The last official flight of the Apollo programme broke several records: for the longest time in lunar orbit, the largest sample brought back from a mission, the longest total moonwalks, and the longest Moon landing itself. It was also the first time that NASA had launched a manned spacecraft at night. Sadly, it also currently holds the dubious honour of being the last time humans visited the Moon.

Observer's Guide to the Apollo Landing Sites

Gaze upon the lunar surface tonight and you'll see where man, rovers and landers stepped onto another world

Written by Stuart Atkinson

Today there is a lot of excited talk about going to the Moon. Again. Shining brightly in our sky it calls to us like a celestial siren, just as it always has done. NASA is still debating whether it should send astronauts straight to Mars, bypassing the Moon altogether, or only go to Mars after a number of successful precursor missions to Earth's natural satellite. Meanwhile, the European Space Agency is looking to the Moon as the potential site of a scientific outpost, where different nations could work together in a 'lunar village', much like international scientists now work in Antarctica. Private companies are also planning to mine the Moon for resources, and there's even a competition to land robot rovers on the Moon and have them send live video back to Earth.

With all this going on it's important to remember we've already been to the Moon. True, it happened a long time ago but the Apollo missions were a spectacular success, and represented a golden age of exploration. It was a time when enormous rockets, gleaming white, thundered into the sky, roaring like dragons, carrying brave explorers across the gulf of space, travelling much further than we could possibly go today. Between 1969 and 1972 six Apollo missions took teams of three astronauts across a quarter of a million miles of space to the Moon, set two of them down on its surface, and brought them all home safely again. A seventh mission, Apollo 13, famously failed to land on the Moon, but the astronauts survived a flight around the Moon. Today those daring missions are as fascinating as ever.

Many people have asked why astronomers don't turn the Hubble telescope towards the Moon, to take photos of the Apollo spacecraft. But not even the Hubble could see a four-metre wide Apollo spacecraft on the Moon. Hubble is essentially a light bucket, designed to collect the faint, ghostly light of faraway galaxies, nebulae and planets. It can't zoom in on things in its own backyard.

To see Apollo hardware you have to go to the Moon, and then either land next to the actual spacecraft, as the rovers might do later this year or next, or look down on them from orbit. The Lunar Reconnaissance Orbiter (LRO), has done just that, and has taken amazing images of the Apollo landing sites from orbit showing not just the spacecraft themselves, but the lunar rovers parked where they were left, and even the trails of bootprints left in the lunar dust by the explorers.

So, if you were hoping to see Apollo hardware on the Moon through your telescope, you've no chance, sadly. However, you can see the Apollo landing sites if your telescope is good enough — and we're going to tell you how, and where, to find them.

First, you need to know the general areas of the landing sites, and the key to doing that is to think of the Moon as the face of a clock, with 12 o'clock at the top and 6 o'clock at the bottom. You can then find the rough areas of each mission's landing site quite easily, using the charts included in this guide.

Having found the general areas of the landing sites, you can then zoom in on those to pin-down the actual landing sites. You do this by looking for certain features the landing sites were close to, such as a large crater or a valley. Again the charts will help you. Note: the charts are oriented correctly for the 'upside down' view seen through most telescopes.

Apollo 11

Telescope with magnification of 50x or more

Mare Tranquillitatis (Sea of Tranquility)

Between first quarter and full

Finding Apollo 11's landing site where Neil Armstrong took his "one small step" off the Eagle's ladder is quite easy. Just find the large crater Theophilus and put it at the top of your field of view. You'll see an obvious 'promontory' of bright ground beneath the crater, jutting out into the darker lava sea. The 'Tranquility Base' is just beneath this striking feature.

Apollo 12

Telescope with magnification of 50x or more

Oceanus Procellarum (Ocean of Storms), close to crater Copernicus

Between full and last quarter

One of the Moon's most impressive craters will guide you towards the Apollo 12 landing site in the Ocean of Storms. Just find the huge crater Copernicus and place it at the bottom of your inverted field of view. To Copernicus' upper right you'll see the smaller crater Reinhold, and beyond it the crater Lansberg. Apollo 12's landing site lies to the upper left of the 3.1-kilometre-deep Lansberg.

Apollo 14

Telescope with magnification of 50x or more

Fra Mauro, close to crater Ptolemaeus

Between first quarter and full Moon

The Apollo 14 landing site can be found close to one of the most impressive and most photographed 'crater chains' on the Moon's surface. Once you have found craters Arzachel, Alphonsus and Ptolemaeus, jump across to the right of Ptolemaeus where you will find the smaller ring-like crater Parry. The Apollo 14 landing site is just to the lower right of this crater.

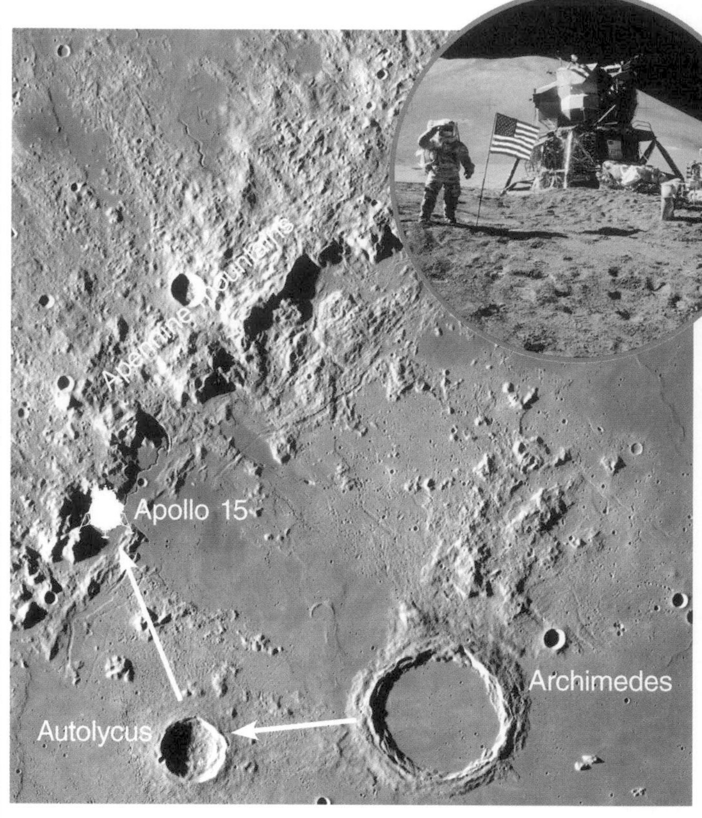

Apollo 15

Telescope with magnification of 100x or more

Close to Hadley Rille, in the Apennine mountains

Between first quarter and full Moon

The lunar module Falcon touched down in July 1971 in the most stunning location any Apollo mission visited — close to a meandering valley in the shadow of the Apennine mountains. To find it, look for the break in the curve of the mountains, to the left of the crater Archimedes, past Autolycus and Aristillus. Apollo 15 landed above and to the left of these craters, in the foothills of the mountains.

Apollo 16

Telescope with magnification of 100x or more

The Descartes Highlands, close to the crater Kant

First quarter to full

The landing site of Apollo 16's lunar module 'Orion' is probably the most challenging to find. If you place the crater Theophilus to the left of your eyepiece's field of view you'll see a smaller, sharper-rimmed crater to its right. This is Kant, and Apollo 16 set down in the rugged highlands to its lower right.landing site.

Apollo 17

Telescope with magnification of 100x or more

Taurus-Littrow Valley

First quarter to full

The final Apollo mission in December 1972 saw the lunar module Challenger land in a notch-like 'bay' on the southern shore of the Sea of Serenity. To find it, put the shallow crater Posidonius at the bottom of your field of view. Follow the shoreline 'up' past the semi-circular Le Monnier bay. Continue upwards and you'll find the Apollo 17 landing site.

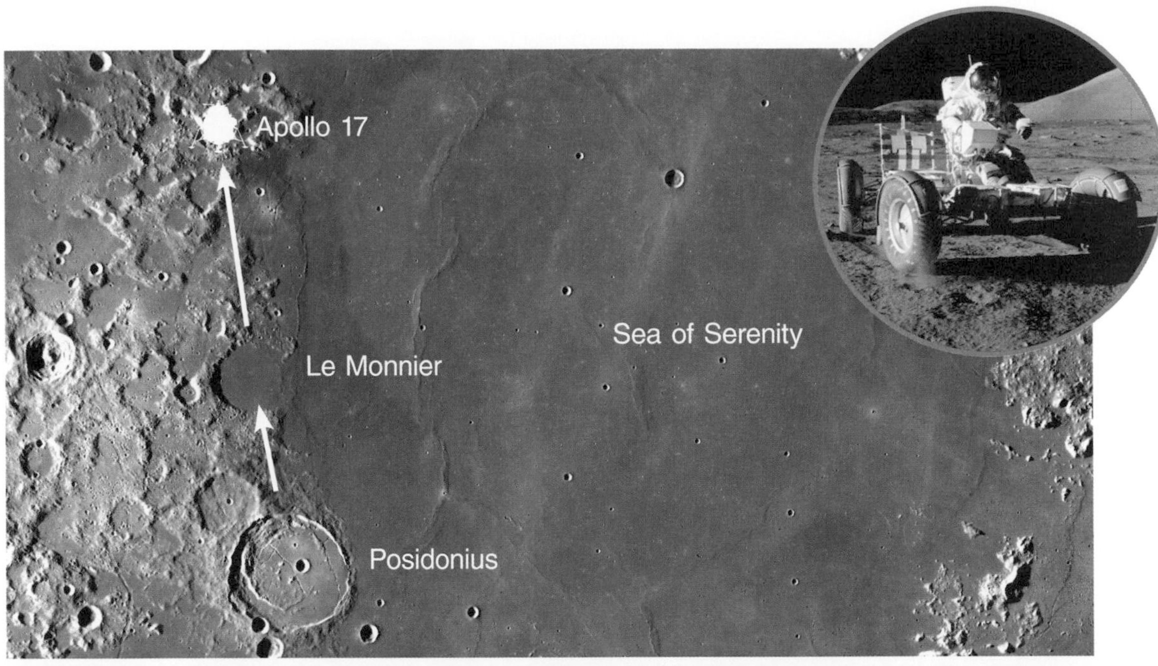

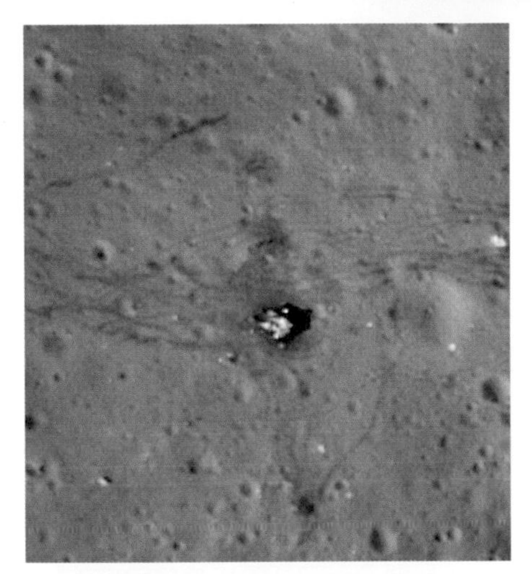

This image shows the Challenger Descent Stage of the Apollo 17 mission as well as buggy tracks and footprints

The Apollo 13 mission was aborted after an oxygen tank ruptured. Here we can see the impact site of its booster

Using the Moon as a clock face

Locate the rough locations of the Apollo landing sites using this simple trick

© Alamy; Detlev Van Ravenswaay; Science Photo Library; NASA; Goddard; Arizona State University; Gregory H. Revera

Dress rehearsal for the Moon

The crew of Apollo 10 performed every aspect of an actual mission to the lunar surface except the landing itself

The fourth mission of NASA's Apollo program in the short span of seven months, Apollo 10 lifted off from the Kennedy Space Center's Launch Complex 39B on the morning of 18 May 1969. Its objective was clear: to execute every aspect of a lunar landing mission except the actual landing on the Moon itself. That defining moment would, hopefully, occur weeks later with Apollo 11. However, unless Apollo 10's dress rehearsal was successful, NASA's goal of placing a man on the Moon before the end of the 1960s would be in serious jeopardy.

Apollo 10 was the first lunar mission involving an entire Apollo spacecraft configured for a landing. The command/service module was a two-part vehicle, its cone-shaped command module used as a control station and crew compartment. The cylindrical service module, to the rear of the command module, contained oxygen, hydrogen, fuel, and propulsion and manoeuvre systems. The two-stage lunar module housed a lower descent stage containing the power plant for the Moon landing. Its construction included four aluminium alloy legs for support on the lunar surface, a ladder for astronaut ingress and egress, and storage space. The descent stage also provided a launch platform for the cylindrical ascent stage, where the crewmen would work while on the Moon's surface prior to lifting off, returning to lunar orbit, and docking with the command/service module.

During the eight-day mission, Apollo 10 set the record for highest speed ever achieved by a manned vehicle at 39,897 kilometres per hour (24,791 miles per hour) while returning to Earth, and achieved the greatest distance humans have ever travelled from home, 408,950 kilometres (220,820 nautical miles) from the crew's houses and NASA mission control in Houston, Texas.

Mission commander Thomas Stafford, an Air Force officer, was a NASA veteran of Gemini 6 and Gemini 9. He was later the mission commander of the Apollo-Soyuz Test Project, a joint venture with the Soviet space program. Lunar Module pilot Eugene Cernan, a Navy officer, flew with Stafford aboard Gemini 9 and later commanded Apollo 17, becoming the eleventh man to walk on the Moon. Command Module pilot John Young, a

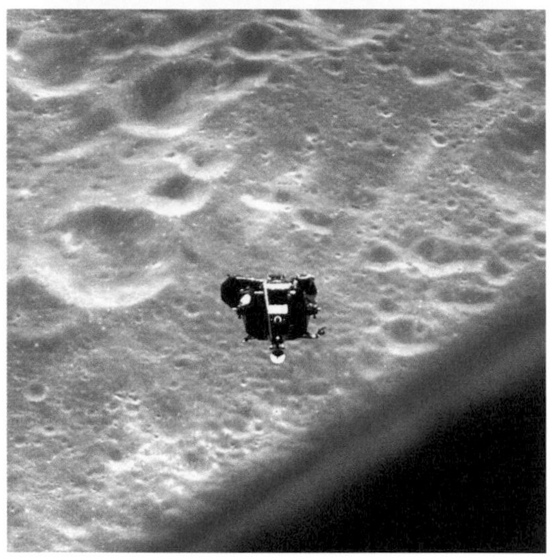

Navy officer, had flown in Gemini 3 and Gemini 10. He later flew in Apollo 16, becoming the ninth man to walk on the Moon, and commanded two Space Shuttle missions. During Apollo 10, Young became the first person to fly solo around the Moon.

The crew nicknamed its command/service and lunar modules Charlie Brown and Snoopy after popular characters from the Peanuts comic strip, and cartoonist Charles Schulz created artwork for the project. The intent was to add a bit of familiarity to the Moonshot, but NASA officials considered the nicknames undignified. Nevertheless, the idea of whimsical names persisted with later Apollo missions.

A total of 12 television broadcasts were originally planned during the flight, the first to transmit colour images from space. The initial live transmission began three hours after launch, and the cameras provided stunning colour footage of the Earth and the surface of the Moon. By the end of the mission, 19 transmissions totalling nearly six hours had occurred, also offering viewers clear depictions of life aboard the spacecraft.

Once aloft, Apollo 10 completed one and a half revolutions around the Earth before igniting the S-IVB booster stage of its Saturn V rocket, gaining sufficient velocity to escape Earth's gravitational pull. Three days later the spacecraft reached a lunar orbit 111.12 kilometres by 314.84 kilometres (60 by 170 nautical miles) above the lunar surface.

On 22 May, Stafford and Cernan moved into the lunar module, detaching from the command/service module to execute a simulated lunar landing. They proceeded to a temporary orbit of 107.34 kilometres by 115.07 kilometres (66.7 miles by 71.5 miles). The descent engine fired for 27.4 seconds, and the lunar module manoeuvred to an orbit of 15.61 kilometres by 113.45 kilometres (9.7 miles by 70.5 miles). The astronauts then surveyed the proposed lunar landing site in the Sea of Tranquillity while their pre-landing tests were performed.

NASA officials had been concerned that Stafford and Cernan might actually go rogue and land the lunar module themselves. Cernan was later quoted as saying, "A lot of people thought about the kind of people we were: 'Don't give those guys an opportunity to land, 'cause they might!'" To eliminate the worry, the tanks in the ascent module were deliberately shorted of fuel. If the Apollo 10 astronauts had landed on the Moon, they could not have returned to the command/service module, where Young was flying alone.

While the command/service and lunar modules were separated, all three astronauts picked up an eerie whistling sound that they went on to describe as "space music". While Young correctly identified the source as radio interference between the two modules, later disclosure of the event brought about some implausible speculation among observers that some kind of extraterrestrial communication had taken place.

This portion of the mission narrowly averted disaster when the time came to jettison the descent stage and return to the command/service module. The descent stage separated on the second attempt, and immediately the ascent stage experienced violent rolls. Cernan shouted an expletive that was broadcast worldwide as he struggled to bring the ascent stage under control. He counted eight spirals and managed to pull the vehicle out of the spin with little time to spare before a fatal impact with the lunar surface. Post-flight analysis revealed that a single switch had been in the wrong position and caused the near-catastrophe. After reaching the proper orbit, Stafford sighted the command/service module at a distance of 77.25 kilometres (48 miles). The vehicles re-docked successfully on 23 May, and the ascent stage was jettisoned. The crew continued routine activities for the remainder of the mission.

The next day, Apollo 10 headed for home. Splashdown in the Pacific Ocean occurred on May 26, about 6.4 kilometres (4 miles) from the recovery ship, the aircraft carrier USS Princeton. The pathway to the Moon was thoroughly charted.

© NASA

Apollo 11 The inside story

What really happened the day we landed on the Moon

Written by Nick Howes

It's hard for many in 2019 to comprehend that 50 years ago, humankind achieved one of the greatest technical feats of all time. Less than nine years after President Kennedy had set the goal of landing a man on the surface of the Moon and returning him safely to Earth, NASA achieved that most astonishing aim on 20 July 1969.

Those intervening years had been a white-knuckle ride. Beginning with Alan Shepard's 15 minute sub-orbital Mercury flight in 1961, NASA progressed through a series of milestones in their mission to reach the Moon. There was the loss of a Mercury capsule and the near-drowning of its pilot Gus Grissom; John Glenn's re-entry with a retro-rocket still attached to his Friendship 7 capsule; a slew of hugely successful Gemini missions including one that almost span out of control, potentially threatening the life of the astronaut who in 1969 would take that first historic step; and then four fully flown Apollo missions, two in low Earth orbit, two that orbited the Moon and only one to test the full system. NASA had to endure the catastrophic loss of Grissom and his two crew mates, Edward White and Roger Chaffee in 1967 in Apollo 1's tragic fire on the launch pad, but the space agency had resolved to carry on, completely redesigning the lunar command module and carrying out major changes to the lunar landing module (the LEM as it was known) in that short space of time.

Amid triumph and tragedy, on 16 July 1969 NASA was ready to go to the Moon. Yet the trials and tribulations of the previous years were not over and the three-man crew of Apollo 11 — Neil Armstrong, Buzz Aldrin and Michael Collins — were facing one of the most dramatic spaceflights in history.

We recall the historic first words said on the lunar surface, and the elation of the largest TV audience in history at that time when they saw those grainy black and white images from the Moon, but there is so much more to the story of Apollo 11 that may not be as well known.

Their first task, of course, was to leave Earth on top of the mighty Saturn V rocket — the tallest, most powerful rocket ever built. Many astronauts

"A space mission will never be routine... you're putting three humans on top of an enormous amount of high explosive" Gene Kranz

Armstrong waves to well-wishers in the Manned Spacecraft Operations Building as he, Collins and Aldrin prepare to be transported to Launch Complex 39A

This iconic picture shows astronaut Buzz Aldrin's bootprint in the lunar soil

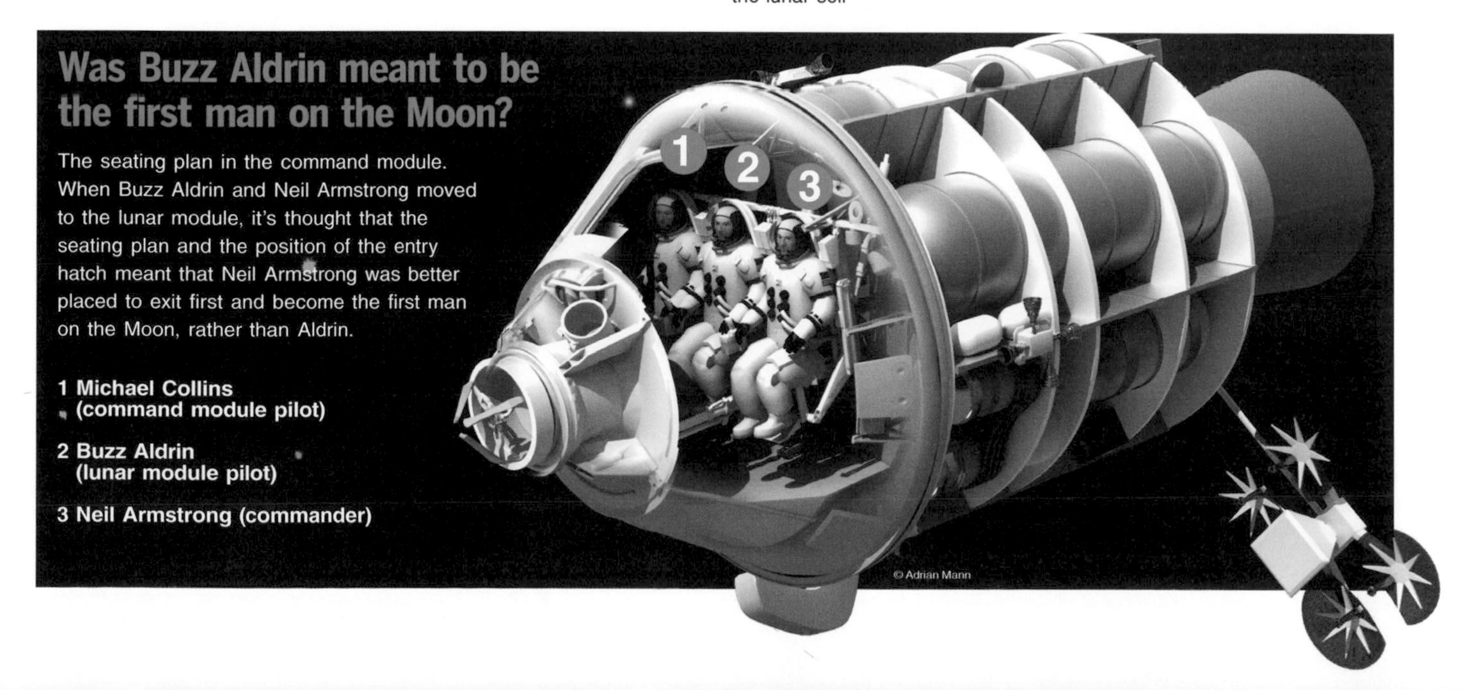

Was Buzz Aldrin meant to be the first man on the Moon?

The seating plan in the command module. When Buzz Aldrin and Neil Armstrong moved to the lunar module, it's thought that the seating plan and the position of the entry hatch meant that Neil Armstrong was better placed to exit first and become the first man on the Moon, rather than Aldrin.

1 Michael Collins (command module pilot)

2 Buzz Aldrin (lunar module pilot)

3 Neil Armstrong (commander)

Buzz Aldrin

After returning to Earth, hardly any shots of the first man on the Moon led Buzz Aldrin to be questioned

It's said that Aldrin was getting Armstrong back by taking no photos of him on the Moon in retribution for the latter getting the honour of being the first to set foot on the Moon. However, and according to Aldrin, he was about to take a picture of Armstrong at the flag ceremomy when President Nixon called, distracting them from the task. "As the sequence of lunar operations evolved, Neil had the camera most of the time, and the majority of the pictures taken on the Moon that include an astronaut are of me," Aldrin states. "It wasn't until we were back on Earth and in the Lunar Receiving Laboratory looking over the pictures that we realised there were few pictures of Neil. My fault perhaps, but we had never simulated this during our training."

Before his death in 2012, Armstrong defended Aldrin, stating: "We didn't spend any time worrying about who took what pictures. It didn't occur to me that it made any difference, as long as they were good... I don't think Buzz had any reason to take my picture, and it never occurred to me that he should."

"When I got back and someone said, 'There's not any of Neil,' I thought, 'What in the hell can I do now?' I felt so bad about that," says Aldrin. "And then to have somebody say that might have been intentional... How do you come up with a nonconfrontational argument against that?"

"Somebody said that [me not taking pictures of Neil] was intentional"

Buzz Aldrin moves toward a position to deploy two components of the Early Apollo Scientific Experiments Package (EASEP) on the surface of the Moon during the Apollo 11 extravehicular activity

The lunar module pilot poses for a photograph beside the deployed United States flag during an Apollo 11 extravehicular activity (EVA) on the lunar surface

Neil Armstrong works at the lunar module in the only photo taken of him on the Moon from the surface

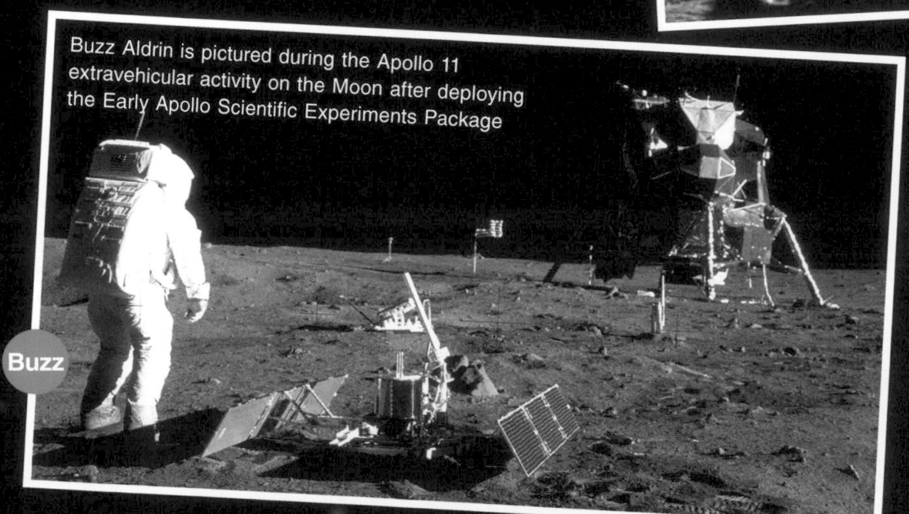

Buzz Aldrin is pictured during the Apollo 11 extravehicular activity on the Moon after deploying the Early Apollo Scientific Experiments Package

Buzz Aldrin walks on the surface of the Moon near the lunar module during the Apollo 11 mission

who were propelled into space by the Saturn V describe it as being a very smooth ride. Neil Armstrong is quoted as saying that while the launch for all those watching on Cocoa Beach or at Cape Canaveral was deafening, the crew could detect a slight increase in background noise, a lot of shaking, and feeling akin to being onboard a large jet aeroplane on take-off. Yet as smooth a ride as it was, being on top of that much rocket fuel was always a dangerous experience.

"A space mission will never be routine because you're putting three humans on top of an enormous amount of high explosive," Gene Kranz, flight director for Apollo 11's lunar landing, told us. If there were any nerves, the astronauts weren't feeling it, according to Buzz Aldrin. "We felt that involved but there were a lot of points to abort the mission short of continuing on something risky."

"There was a degree of seriousness in mission control that I hadn't even seen in training" Gene Kranz

Once in space, the command service module had to rotate and dock with the lunar module, which was embedded in the final S-IVB stage of the Saturn V rocket. After the two spacecraft had mated, onwards they flew to the Moon, leaving the S-IVB stage trailing in space behind them.

Some time later, the crew spotted something strange outside. A light that appeared to be following them. When Michael Collins used the onboard telescope to view it, he couldn't make it out — it looked like a series of ellipses but, when focusing the telescope, it seemed L-shaped, but that could have just been the way sunlight was glinting off it.

Reticent to tell mission control in Houston, Texas, that they were being raced to the Moon by a UFO, the crew cautiously asked where the S-IVB rocket stage was. "A few moments later they came back to us and said it was around 6,000 miles away," recalled Aldrin. "We really didn't think we were looking at something that far away, so we decided to go to sleep and not talk about it any more."

Aldrin doesn't believe it was an alien spaceship, but that it was more likely the Sun reflecting off one of four metal panels that fell away from the rocket stage when they docked with the lunar module.

For almost four days Apollo 11 flew towards the Moon, where Armstrong and Aldrin climbed into the lunar module — the Eagle — and said goodbye to Collins, who was to remain in the command module in orbit around the Moon.

As the Eagle flew around the far side of the Moon, things in mission control were growing tense. "There was a degree of seriousness in mission control that I hadn't even seen in training," said Kranz. "That was when you realised this was the real deal: today, we land on the Moon."

Almost immediately after separating from the command module there were problems. Radio communication with the Eagle was sketchy at best and they were coming up to the point of no return, where the landing could no longer be aborted if something was wrong.

"It was up to me to decide if we had enough information to make the go/no-go [decision] and continue the descent to the Moon," said Kranz. So, five minutes before the powered descent to the lunar surface was due to begin, with radio communication cutting in and out, Kranz asked his flight controllers to give him their go/no-go based on the last frame of data that they saw. They all said "go". And then things turned from bad to nearly catastrophic.

The spacecraft's guidance computer, developed at MIT under the auspices of Charles Draper (the lab at MIT now bears his name) was a 2MHz system that was the first in the world to use integrated circuits. Its fixed memory was an ingeniously designed 'Core Rope', which consisted of a set of small hoops that 'Little Old Ladies' (as it was referred to at the time) along with machines would thread the code either through or around the hoops to give the computer its 1 or 0 value. If the MIT code was threaded incorrectly, the 'programmer' would have to laboriously go through the woven cores and debug it.

The flight controllers erupt into applause as Apollo 11 splashes down in the Pacific Ocean on 24 July 1969, successfully completing the mission

The huge, 363-feet tall Saturn V rocket carries three men towards the Moon from Pad A, Launch Complex 39, Kennedy Space Center on 16 July 1969

After a rehearsal mishap when the Lunar Landing Research Vehicle exploded, Neil Armstrong floats safely to the ground

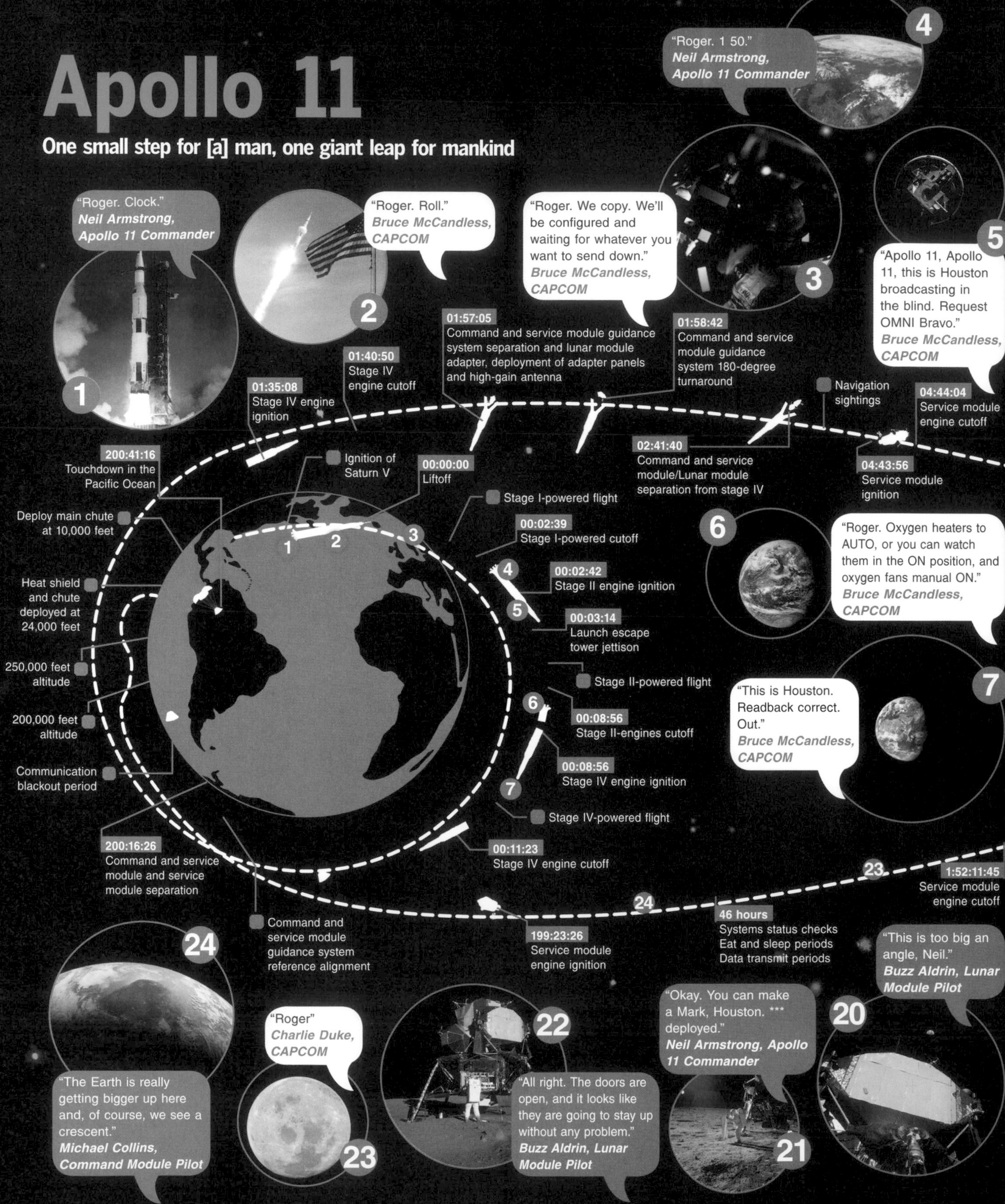
Apollo 11
One small step for [a] man, one giant leap for mankind
1
"Roger. Clock." Neil Armstrong, Apollo 11 Commander
2
"Roger. Roll." Bruce McCandless, CAPCOM
3
"Roger. We copy. We'll be configured and waiting for whatever you want to send down." Bruce McCandless, CAPCOM
4
"Roger. 1 50." Neil Armstrong, Apollo 11 Commander
5
"Apollo 11, Apollo 11, this is Houston broadcasting in the blind. Request OMNI Bravo." Bruce McCandless, CAPCOM
6
"Roger. Oxygen heaters to AUTO, or you can watch them in the ON position, and oxygen fans manual ON." Bruce McCandless, CAPCOM
7
"This is Houston. Readback correct. Out." Bruce McCandless, CAPCOM
01:35:08 Stage IV engine ignition
01:40:50 Stage IV engine cutoff
01:57:05 Command and service module guidance system separation and lunar module adapter, deployment of adapter panels and high-gain antenna
01:58:42 Command and service module guidance system 180-degree turnaround
Navigation sightings
04:44:04 Service module engine cutoff
02:41:40 Command and service module/Lunar module separation from stage IV
04:43:56 Service module ignition
200:41:16 Touchdown in the Pacific Ocean
Ignition of Saturn V
00:00:00 Liftoff
Stage I-powered flight
00:02:39 Stage I-powered cutoff
00:02:42 Stage II engine ignition
00:03:14 Launch escape tower jettison
Stage II-powered flight
00:08:56 Stage II-engines cutoff
00:08:56 Stage IV engine ignition
Stage IV-powered flight
00:11:23 Stage IV engine cutoff
Deploy main chute at 10,000 feet
Heat shield and chute deployed at 24,000 feet
250,000 feet altitude
200,000 feet altitude
Communication blackout period
200:16:26 Command and service module and service module separation
Command and service module guidance system reference alignment
199:23:26 Service module engine ignition
46 hours Systems status checks Eat and sleep periods Data transmit periods
1:52:11:45 Service module engine cutoff
1 2 3 4 5 6 7 23 24
24
"The Earth is really getting bigger up here and, of course, we see a crescent." Michael Collins, Command Module Pilot
23
"Roger" Charlie Duke, CAPCOM
22
"All right. The doors are open, and it looks like they are going to stay up without any problem." Buzz Aldrin, Lunar Module Pilot
21
"Okay. You can make a Mark, Houston. *** deployed." Neil Armstrong, Apollo 11 Commander
20
"This is too big an angle, Neil." Buzz Aldrin, Lunar Module Pilot

9

"Okay, no complaints. I was just curious as to what had happened."
Michael Collins, Command Module Pilot

10

11

"Okay."
Michael Collins, Command Module Pilot

"Apollo 11 is getting its first view of the landing approach. This time we are going over the Taruntius crater, and the pictures and maps brought back by Apollo 8 and 10 have given us a very good preview of what to look at here. It looks very much like the pictures, but like the difference between watching a real football game and watching it on TV. There's no substitute for actually being here."
Neil Armstrong, Apollo 11 Commander

12

"That's a good, reasonable way of describing it. It appears as though it made a difference just sitting back in the tunnel and gazing at all windows; it makes a difference which one you're looking out of. The camera right now is looking out the number 5 window, and it definitely gives a rosier or tanner tinge."
Buzz Aldrin, Lunar Module Pilot

13

"Apollo 11, Houston. Thirty seconds to loss of signal. Both spacecraft looking good going over the hill. Out."
Charlie Duke, CAPCOM

14

"I think you've got a fine looking flying machine there, Eagle, despite the fact you're upside down."
Michael Collins, Command Module Pilot

46 hours
Systems status checks
Eat and sleep periods
Data transmit periods

51:40:59
Service module engine cutoff

51:40:51
Service module ignition

9 hours
Systems status checks
Eat and sleep period
Data transmit period

62:16:57
Service module engine ignition

62:17:01
Service module cutoff

66:17:43
Pilot transfer to lunar module, second orbit

64:04:38
Begin navigation sightings

Lunar descent

70:37:45
Lunar touchdown

70:27:17
Lunar descent engine ignition

Lunar module guidance system and reference alignment

69:29:03
Lunar module descent engine cutoff

Begin lunar module systems activation and checkout

Begin lunar orbit evaluation

Transfer orbit insertion

63:23:27
Service engine cutoff

69:28:31
Lunar module descent engine ignition

Lunar orbit insertion

69:05:32
Command and service module and lunar module separate on third orbit

66:45:53
Commander transfer to lunar module

8

"11, Houston. If that's not the Earth, we're in trouble."
Charlie Duke, CAPCOM

9 hours
Systems status checks
Eat and sleep period
Data transmit period

109:00:04
Command and service module and lunar separate and lunar module jettison

105:19:04
Liftoff

122:11:44
Service module engine ignition

Lunar module ascent

28 hours
Systems status checks
Eat and sleep periods

1:52:11:44
Service module engine ignition

Lunar module ignition

Transfer crew and equipment from lunar module to Command and service module

15

"See you later."
Neil Armstrong, Apollo 11 Commander

Midcourse corrections

Rendezvous maneoures

108:02:14
Command and service module and lunar module initial docking

19

"For those who haven't read the plaque… First there's two hemispheres, one showing each of the two hemispheres of the Earth. Underneath it says "Here Man from the planet Earth first set foot upon the Moon, July 1969 A.D. We came in peace for all mankind." It has the crew members' signatures and the signature of the President of the United States."
Neil Armstrong, Apollo 11 Commander

18

"Contingency sample is in the pocket. My oxygen is 81 per cent. I have no flags, and I'm in minimum flow."
Neil Armstrong, Apollo 11 Commander

17

"The surface is fine and powdery. I can pick it up loosely with my toe. It does adhere in fine layers like powdered charcoal to the sole and sides of my boots. I only go in a small fraction of an inch, but I can see the footprints of my boots and the treads in the fine particles."
Neil Armstrong, Apollo 11 Commander

16

"Roger, Tranquility. We copy you on the ground. You got a bunch of guys about to turn blue. We're breathing again. Thanks a lot."
Charlie Duke, CAPCOM

The Apollo 11 astronauts, left to right, Neil Armstrong, Michael Collins and Edwin 'Buzz' Aldrin inside the Mobile Quarantine Facility are greeted by President Nixon on 24 July 1969

An inside view of the Apollo 11 lunar module shows astronaut Buzz Aldrin during the lunar landing mission in an image taken by Neil Armstrong

Mission control loses contact with Apollo 11

Alarms, loss of communication and system failures plagued the first mission to land on the Moon

03:04:15:47 "Apollo 11, Apollo 11, this is Houston. Do you read? Over." *Bruce McCandless, CAPCOM*

03:04:15:59 "Apollo 11, Apollo 11, this is Houston. Do you read? Over." *Bruce McCandless, CAPCOM*

03:04:16:11 "…" *Unidentified crew member, Apollo 11*

03:04:16:59 "Houston, Apollo 11. Over." *Unidentified crew member, Apollo 11*

03:04:17:00 "Apollo 11, Apollo 11, this is Houston. We are reading you weakly. Go ahead. Over." *Bruce McCandless, CAPCOM*

03:04:19:32 "Apollo 11, this is Houston. Are you in the process of acquiring data on the burn? Over." *Bruce McCandless, CAPCOM*

03:04:21:37 "Apollo 11, Apollo 11, this is Houston. How do you read?" *Bruce McCandless, CAPCOM*

03:04:21:43 "Reading you loud and clear, Houston. How us?" *Neil Armstrong, Apollo 11 Commander*

© Adrian Mann

When the crew were approaching the Moon for the landing, various alarms were triggered by the computer. "Whatever information we were looking at [disappeared] and instead it gave us the code number of the alarm," said Aldrin. "It was disturbing and distracting and we didn't know what it meant."

The 1201 and 1202 alarms were obscure codes (and in effect the same error) that flashed up as Armstrong manually attempted to bring the lunar module down. Nobody seemed to know what the codes meant, except for two men: Jack Garman, a NASA computer engineer who had come across the codes before during a practice run, and Steve Bales, who was the Apollo guidance officer. The alarms were being caused by a problem with the landing radar that was stealing precious computing cycles, and the throttle control algorithm was barely working. The computer's 72kb of memory, barely enough to write a sentence in a modern word processor, was struggling as commands into it overflowed. Garman knew that it was safe to continue and allow the computer to handle matters. Its priority scheduling routines, which have formed some of the basis of a lot of modern code, were dumping lower priority tasks in favour of the ones critical to the lunar landing.

As the Eagle approached the surface on automatic, Armstrong and Aldrin realised that the scenery outside of the window didn't look familiar to them. "I think we may be a little long," commented Armstrong, referring to the Eagle having overshot its planned landing site. Looming ahead of them inside a crater was a dangerous-looking boulder field, and coming down on any of those giant rocks the size of houses would have damaged or perhaps even destroyed the Eagle. Armstrong took manual control, using the thrusters to take the Eagle over the boulder field. But now fuel was running low and there was no turning back. Armstrong had to land the Eagle — somewhere, within minutes — or they would be out of fuel and crash.

"We'd never been this close in training," said Kranz. "We started a stopwatch running, with a controller calling off seconds of fuel remaining."

If things were tense in mission control, onboard the Eagle Armstrong and Aldrin had everything under control. With only 13 seconds of fuel left, Apollo 11 made its safe landing in the Sea of Tranquillity. History had been made. "Houston, Tranquility Base here," Armstrong radioed home. "The Eagle has landed."

In private, Aldrin took out a small cup, some wine and bread and said Holy Communion. The wine, under one-sixth Earth gravity, apparently curled up in the cup. After reading a section of the Gospel of St John, Aldrin said a few words, with Armstrong respectfully just looking on. NASA had been threatened with legal action by Madalyn O'Hair, an atheist, after the crew of Apollo 8 had read from the Book of Genesis, so Aldrin's heartfelt ceremony never made it to the airwaves. Aldrin, though, has always been content in the thought that the first food and drink consumed on the lunar surface were communion items.

The original plan had been for the crew to get some sleep after the landing, but with that much adrenaline pumping through their veins that was never going to happen. So at 2.39am on the morning of 21 July, Armstrong made his way through the hatch and down the ladder before stepping foot for the first time on the surface of the Moon and saying those now-immortal words, "That's one small step for [a] man, one giant leap for mankind."

After exiting the lunar module, Armstrong and Aldrin only had a few hours to not only collect precious rock samples, but also deploy a series of experiments on the lunar surface. Solar wind experiments, a laser retro-reflector that is still used to this day to measure the Earth-Moon distance, seismometers and more were all deployed. Armstrong is cited as saying he felt like a five-year-old in a candy store, with not enough time to do all the things he wanted to.

Standing on the Moon must have been an incredible experience. Aldrin described the scene around him as one of "magnificent desolation", adding that, "You could look at the horizon and see very clearly because there was no atmosphere, there was no haze or anything."

As Armstrong walked around setting up instruments and picking up rocks, Aldrin hopped around on the surface, testing what the best way to move about in the low gravity was. Most of the pictures taken during the landing are of Aldrin on the surface; barely half a dozen show Armstrong, and none clearly. That's because Armstrong had the camera for most of the Moon walk.

While on the surface, the crew also had terrific problems with the American flag. It had a telescoping boom arm to hold it out in lieu of any wind to hold it up. The two crew wrestled to get the boom arm to extend fully, but it would not, so the flag had a small kink in it. They also found that it was almost impossible to get the flag pole to go deep enough into the ground and, in the end, they only just managed to get it to stay upright. Both of the crew worried it would fall over live on TV, and probably as President Nixon was on the phone to them. But it remained upright during the broadcasts and telephone calls.

After collecting their rocks and clambering back into the lunar module, the crew took off their boots and backpacks, and began to throw anything not of vital importance back on to the lunar surface. This included urine bags, empty food packs, empty cameras and so on. But to the crew, they were just getting in the way and not needed.

There was time for one final crisis. The interior of the lunar module was cramped and, moving around in their bulky spacesuits, one of the astronauts had knocked out the switch for the circuit breaker that fired the ascent rocket that would take them home.

This was a real bottleneck moment for the mission. "If for some reason the ascent engine didn't work, there was no way to rescue the crew," said Kranz. Armstrong and Aldrin would be stranded on the Moon. The concern was so serious that President Nixon had a speech prepared, while mission control would close down communications with Armstrong and Aldrin after a clergyman had "condemned their souls to the deepest of the deep". Without that circuit breaker the crew were facing that lonely fate, but their training would not have allowed them to give up. "Rather than worry about things like that, we'd face them when the time came and we'd work as hard as we could to fix the problem until our oxygen ran out," said Aldrin. In the end, the solution was remarkably simple: jabbing the end of a pen into the slot where the broken switch had been, Aldrin was able to push the circuit breaker in. The ascent rocket fired and the two Moon-walkers were on their way home, via a rendezvous with Michael Collins in the command module. As the Eagle took off, the flag finally did blow over, and to this day it lays flattened, bleached out by solar radiation.

Almost 50 years since that first successful landing on the Moon, stories still come out, not just from the thoughts of the crew, but also the almost 400,000 others who worked on the mission, from 'the guy sweeping the floor' at Cape Canaveral, to the flight directors and flight controllers still, without whom the historic landing may never have happened. With our return to the Moon still some way off, these stories are all we have for now.

How to...

Use a felt-tipped pen to escape from the Moon

After a circuit breaker switch broke off in all the too-ing and fro-ing in the cramped environment of the lunar module, Buzz Aldrin had to improvise in order to escape the Moon.

01 Astronauts locate broken circuit breaker switch

Neil Armstrong and Buzz Aldrin were gathering themselves into the landing module to start the return back to Earth when Aldrin noticed something lying on the floor — the circuit breaker switch had gotten bumped and had broken off.

02 Aldrin and Armstrong alert mission control

This switch needed to activate the ascent engine to lift them off the Moon. Telling mission control, they tried unsuccessfully to catch some sleep but, by the following morning, NASA had no idea what to do with Aldrin forced to come up with a solution.

03 Saved by a felt-tipped pen

Since the circuit was electrical, sticking his finger or anything metal in wasn't possible. Instead, Aldrin found a felt-tipped pen in his shirt and inserted it into the opening where the circuit breaker switch should have been. He moved the countdown procedure up by a couple of hours.

04 Lift off!

The circuit breaker held, allowing both Aldrin and Armstrong to lift off from the surface of the Moon and intercept Michael Collins, who was in orbit around the Moon.

© Ed Crooks

© NASA

Houston, we've had a problem

Planned as NASA's third lunar landing, Apollo 13 gripped the world's attention for all the wrong reasons, as a flight to the Moon turned into a battle for survival

On 11 April 1970, as Apollo 13 blasted clear of Cape Canaveral right on schedule, none of those on board could have foreseen the struggle for survival they were soon to encounter. The crew included two freshman astronauts — Command Module pilot Jack Swigert and Lunar Module pilot Fred Haise — but was led by one of NASA's most experienced spacefarers, Gemini and Apollo 8 veteran Jim Lovell (Lovell and Haise had been backup crew for the Apollo 11 mission, while Swigert was a late replacement for Ken Mattingly, who had been grounded as an infection risk after one of his children contracted rubella).

The first two days of their cruise towards the Moon were routine, but 56 hours into the flight, a routine request to stir the service module's fuel tanks rapidly spiralled into a crisis. As Swigert triggered the stirring motor, a loud bang echoed through the craft and warning lights flashed to indicate that one of the module's power circuits was rapidly draining. Swigert and Lovell reported back to mission control with typical understatement: "Okay Houston, we've had a problem here."

Now the spacecraft began shaking from side to side, and as Lovell struggled to stabilise it, he spotted a jet of gas escaping into space. On-board gauges and telemetry signals received at Houston showed one of the service module's two oxygen tanks as empty, and two of the three batteries designed to power the command and service modules (CSM) throughout the mission were now flat. Even worse, pressure in the second oxygen tank was falling.

Now, the problem became clear — the spacecraft oxygen tanks provided not just fuel and air, but were also linked to a fuel cell that charged the batteries. An explosion (later traced to poorly insulated wiring) had ruptured the system, but the tanks were still pumping oxygen to it.

With Apollo 13 some 330,000 kilometres (205,000 miles) from Earth and still Moon-bound, the crew and staff at mission control, led by flight director Gene Kranz, had to think fast. After shutting down the fuel cell to preserve the remaining oxygen, their first thought was to draw power from the independent systems on the Lunar Module (LM) Aquarius, but this idea was soon abandoned, as the demands on the LM's limited batteries would be too high. Instead, the crew were ordered to use Aquarius as a 'lifeboat', transferring supplies into the cramped vehicle (only intended for two astronauts), before shutting down the CSM systems completely to preserve them for return to Earth, and locking themselves in.

Within three hours, the immediate crisis was over, but the struggle to get the crew safely back to Earth was just beginning. NASA's contingency plans to abort a mission in this phase called for jettisoning the LM and also required a fully fuelled CSM, so were obviously out of the question.

Instead, Kranz and his team realised the only option was to swing the entire spacecraft around the far side of the Moon, using the LM's small engines to enter a return trajectory. The timing of these engine burns would be critical — 30.7 seconds was needed on lunar approach to put the spacecraft into a 'free return' trajectory (where the Moon's gravity would effectively swing the spacecraft around and hurl it back towards Earth), and then a longer burn during return would speed up re-entry by ten hours (so that splashdown would occur in the Pacific rather than the Indian Ocean). As the crippled spacecraft swung around the far side of the Moon on 15 April, and communications with Earth were temporarily cut off, its crew set an unwanted new record as the furthest humans from Earth, some 400,171 kilometres (248,655 miles) away.

Sealed into the cramped Lunar Module, survival for the crew now became the overriding priority; all four ground-control shifts at Houston were drafted in to look at various aspects of the problem. Oxygen supplies were sufficient even with three men on board, but water was limited, and power consumption had to be reduced, so television transmissions were abandoned and even radio communications scaled back. A critical danger, however, was the buildup of toxic carbon dioxide — both elements of the spacecraft used canisters loaded with a chemical called lithium hydroxide to 'scrub' the excess CO2 from the air, but the LM's supply was being rapidly used up, and even though the astronauts had brought over extra canisters from the CSM, they were not compatible. Working with a list of available materials on the LM, engineers at Houston came up with instructions for the crew to make an improvised adapter, nicknamed the 'mailbox', using a spacesuit connecting hose.

As the spacecraft neared Earth, a final set of challenges awaited. Most critical was the need to power up the CSM from its shutdown state without causing further damage. No one had thought that such a procedure would ever have to be done in space, and the grounded Mattingly, along with flight controller John Aaron and others, worked feverishly to develop a safe procedure.

On 17 April, with millions around the world listening in to commentary on their every move, the astronauts re-entered the command module Odyssey and brought it back to life. After abandoning the damaged service module, one last risky procedure involved separating from the trusty Aquarius and pushing it away by forcing air into the connecting tunnel between the two modules.

Having avoided the risks of a collision during re-entry, the cone-shaped command module plunged back into the atmosphere. With no way of knowing whether the explosion had damaged the heat shields or descent parachute system, Houston and the world held their breath — and tension worsened as the usual radio blackout lengthened from an expected four and a half minutes to six. Finally, to everyone's relief, Swigert's voice emerged over the crackling radio.

Nine minutes later, Odyssey splashed down, within 6.5 kilometres (4 miles) of the recovery ship USS Iwo Jima, and the celebrations could begin.

Apollo 13 flight directors celebrate the successful splashdown of the Odyssey after the damaged craft's harrowing flight

Swigert aboard Apollo 13 on its return journey to Earth, with some of the improvised hoses used to filter the astronauts' air supply

© NASA

Apollo 14's Moon landing

In February 1971, three rookies landed on the Moon's surface and played lunar golf

When Apollo 14 launched on 31 January 1971, it was not without trepidation. The previous year, the third lunar landing mission — Apollo 13 — had failed to make it to the Moon when an oxygen tank exploded. The crew members survived but serious questions were being asked about the viability of manned missions.

Apollo 14 didn't get off to the best of starts either. Clouds and rain forced a delay of 40 minutes and two seconds, while commander Alan Shepard, command module pilot Stuart Roosa and lunar module pilot, Edgar Mitchell, had problems with the docking latches. But, they finally landed on the Moon on 5 February 1971 in the Fra Mauro formation, with Shepard and Mitchell spending 33.5 hours on the lunar surface. Two extravehicular activities traversed 3.5 kilometres (2.2 miles) over 13 locations, carrying out ten experiments over nine hours. And, although an attempt to tow a two-wheeled trolley full of tools and cameras 1.6 kilometres (one mile) up the steep slopes to the rim of Cone Crater was abandoned, the mission was deemed an overall success. For the trio, it was a particular triumph, as they had been dubbed "the three rookies" due to their lack of spaceflight time and experience.

Indeed, only Shepard had flown before — which was in 1961, as the first American in space. To celebrate, he had his own moment of glory. Just before the crew readied for home, Shepard grabbed a six-iron and hit some golf balls far into the distance. It was, it has to be said, the ultimate Moon shot.

Several experiments were carried out on the Moon and many surface and orbital images were taken

During their Apollo 14 mission to the Moon, Alan Shepard and Edgar Mitchell spent a total of 33.5 hours on the lunar surface

© NASA/Steven Pidcock

Apollo 15

With a crew consisting of David Scott, James Irwin and Alfred Worden, Apollo 15 was a mission that set a number of new records for crewed spaceflight. With the heaviest payload in a lunar orbit, it was the longest crewed lunar mission and the longest Apollo mission at 295 hours. Launching on 26 July 1971, Scott and Irwin spent an astonishing 18 hours and 37 minutes exploring, travelling 17.5 miles in the first ever car on the Moon. Samples were also takes from approximately three metres (ten feet) below the Moon's surface. It was an almost-perfect mission, with the only error being one parachute only partially inflating in the final stages of descent.

© NASA/Steven Pidcock

An interview with...

Astronaut Al Worden

The Apollo 15 Command Module Pilot talks of how the mission made history, why we shouldn't return to the Moon and his enthusiasm for a telescope on the lunar farside

Interviewed by Nick Howes

You've spoken a lot in your book *Falling to Earth* about your love of the UK. Can you talk about your time as a test pilot at the Empire Test Pilot School, where other alumni included Tim Peake?

I attended the Empire Test Pilot School (ETPS) in Farnborough, England, in 1964. I had just completed my time in graduate school at the University of Michigan, and was asked by senior officers to apply for the Air Force Advanced Research Pilot School (AARPS) at Edwards Air Force Base in California, US. While at Michigan I was the ops officer for all the pilots going to college, so I got to know all of them quite well. I even did the flight checks for all pilots, so I really had a double duty while at school.

As a result I was selected for AARPS, but then reassigned to ETPS because of my academic background and the possibility that I would do better in the school than other US pilots had done previously. So I happily went to England with my family. Bill Pogue (who later went on to fly on Skylab), had just completed the course, so he met us at the airport and went with us to London for the first night. It was an exciting time as we had never been out of the US until then.

My time at ETPS started in December 1964 and ended in November 1965. The first few months were mostly classroom work as the weather conditions were not suitable for flying. However, in March we started flying, and before long I was assigned flight duties in 12 or 13 different aircraft. This was a big change from the USAF [United States Air Force] where you could only fly one type of aircraft at a time. I found many differences in the whole philosophy of flying while at Farnborough. We were instructed to make all measurements manually rather than have sensors to do the job. It was a very good and logical method because it forced us to think about the airplane as we were flying it. I learned to fly everything from a Chipmunk to a Viscount while at the school. I think the best part for me was inverted spins in a Hawker Hunter, which, by the way, was a magnificent aircraft.

> "I never had the astronaut programme in my list of things I wanted to do, because I did not think they would have another selection"

I have just recently had the chance to revisit Farnborough for pretty much the first time since 1964. What a change! Back in 1964 we had to keep our night flying up, but they had no runway lights installed. So we flew at night and landed using World War II smudge pots to light the runway. Let's just say, in my day, it was very exciting but a little dangerous when flying jet aircraft.

Lunar Module Pilot James Irwin salutes the United States flag on the Moon on 1 August 1971

So, the record-setting test pilot Chuck Yeager asked you to come back to the US. Did you feel honoured to be asked to teach other pilots?

Yeager was an icon for pilots, being probably the greatest stick and rudder pilot alive, next to Bob Hoover. He was the first person to break the sound barrier in the Bell X-1, and did so while flying with a broken arm. But he was really cool and unafraid, so he did the impossible. I was happy that he wanted me to come back to Edwards to teach. I had the college background that he did not have and he wanted me to write and

Interview Bio

Al Worden

One of just 24 people to have flown to the Moon, Worden was the Command Module Pilot (CMP) for the Apollo 15 mission of 1971. During his NASA career, Worden has logged over 4,000 hours of flying time, including 2,500 hours in jets. He served as a member of the astronaut support crew for Apollo 9 and as back-up CMP for Apollo 12. On top of his many achievements, Worden was listed in the Guinness World Records as the 'Most isolated human being' during his time alone in the Command Module Endeavour.

teach several courses in the advanced section. I never had much interaction with Yeager at the school, but I worked closely with the deputy commandant and the chief of academics to build a programme to teach the things needed for space travel, such as trajectories and free fall (not zero gravity as many believe).

What did it feel like to join the 19 astronauts selected in 1966 for the Apollo programme out of 830 applicants?

I have to say that was one of the high points of my life. Getting a call from Deke Slayton to invite me to join the astronaut programme was overwhelming. I was overjoyed because it had been such a long voyage to get to that point. I never had the astronaut programme in my list of things I wanted to do, because I did not think they would have another selection. However, I was in the right place and had all the credentials they wanted, so I was very optimistic. I had decided that I would be the best test pilot I could be, but then when I had all the things necessary for that, I found they were the same things that NASA was looking for. I didn't hesitate for one second in accepting the invitation.

In your class you had people of the calibre of Jim Irwin (Apollo 15), Charlie Duke (Apollo 16), Stuart Roosa (Apollo 14), Ed Mitchell (Apollo 14) and Fred Haise (Apollo 13). Who do you think excelled from that fifth group of astronauts?

It is hard to say who excelled in my class, because we had some very outstanding guys who were selected. It was, after all, for the Apollo programme, and we were very conscious of what our role would be. Ed Mitchell was clearly the most intelligent, and Joe Engle was clearly the best pilot. But the sum total of flying, academic and personal traits worked out differently; it was a combination of things that put everyone in my class in order.

After our classroom studies, we were asked to rate everyone in the class from one to 18. The flight assignments, which were under the control of Slayton, were then made based on that order. But if you asked me directly which one I thought of as the best, I would have to say Fred Haise. He was the ultimate astronaut.

How do you feel about the Command Module Pilot (CMP) being a largely unsung role compared to the Lunar Module Pilot (LMP) and the Commander?

I have made peace with the idea that the media locked on to those who walked on the Moon, and ignored those of us who stayed in orbit. However, what many fail to understand is that we were the glue that made the flights possible. We did 90 per cent of the flying and navigation, and we were tasked with picking up the others after the surface trip no matter how they got into lunar orbit. If they came off on a crazy trajectory, we would have to go get them, even if it meant that we would not be able to come home.

I think that the media promoted the Moon-walkers because that aspect of the flight was more visually spectacular. Being in orbit was not. The interesting thing was that the CMP had a much bigger role in the total flight than the LMP, who was essentially a flight engineer watching the instrumentation. They did not get the chance to actually fly anything, but they did walk on the Moon, so they became more important to the media.

You are quoted as having felt great about getting most of the flying time on the Apollo 15 mission in comparison to the LMP or Commander — do you still feel this way?

I did get most of the "flying time" on our flight and I took command of the Command Module and Lunar Module after we got to Earth orbit. I first had to extract the Lunar Module from the S-IVB [a stage on the Saturn V rocket], then put us on track to the Moon. I did the navigation and the piloting the entire time of flight.

The Commander [Dave Scott] basically flew the Lunar Module from lunar orbit to the lunar surface and then back to orbit after they had finished their work on the Moon. I actually navigated us back to Earth on my own, without updates from Mission Control, to validate that it was definitely possible to return to Earth without any communications from Mission Control — another very complex element of Apollo 15.

"I also believe that future flights to deep space will require the same kind of training we had because there will probably — initially — not be the luxury of carrying scientists aboard"

Commander Dave Scott leads LMP Jim Irwin and CMP Al Worden to the transfer van ready for the drive to the launch pad

Apollo 15 was regarded as the first of the truly scientific Apollo missions, with extensive training for all the crew in geology. Tell us about your training with the almost legendary Farouk El-Baz. Do you think your education in this area was as good as Dave Scott's and Jim Irwin's?

I had a fabulous time with Farouk while I was learning lunar geology. He was an excellent teacher and I was eager to learn. But there was a big difference in terms of what we learned as a team.

Scott and Irwin learned about rocks and macrogeology while I was taught about large features and significant events that formed the lunar surface. I looked for evidence of volcano activity and meteor impacts as a clue to how the Moon was formed, and saw remarkable features such as cinder cones, which gave the geologists amazing new areas of research to work on. Scott and Irwin looked at the scattering of rocks on the surface as an indication of the results of volcanoes or impact basins.

It was quite different training and hard to compare. Given all of the other work we were doing in flight and mission training, on top of this we practically would have attained a degree in geology if we'd been at university.

You trained for three years, working 70 hours a week with one ten-day break. Do you think that mindset carries forward to modern astronauts?

Training for a lunar flight is not for those with no patience or discipline. It was a long, difficult time and we were very conscious of what could happen if we were not prepared. So, we spent countless days in training. You never get enough training for a flight like that, and anything we missed could have been a disaster, so we trained for everything over and over until we knew it cold. I always felt that we could recover the time spent in training later on, but we had a timeline to live to and wanted to make sure we were ready. And there was always the threat that something could happen to take us off the flight.

I think that today's astronauts have a very different view of spaceflight. They do not have to train for both the flying and the science, with separate disciplines for mission specialists and how the transfers to the International

The Apollo 15 crew stand with the subsatellite they would release into orbit

Space Station (ISS) take place, so there is much less time spent getting ready. Most of the astronauts, without wishing to denigrate their work, are passengers — even those who go to the ISS.

I also believe that future flights to deep space will require the same kind of training we had because there will probably — initially — not be the luxury of carrying scientists aboard. Don't forget that the only 'scientist' (although many of us had master's degrees and/or doctorates in some aspect of science or engineering) to go to the Moon was Jack [Schmitt] on Apollo 17.

Did the launch of Apollo 15 meet your expectations from the training?

The launch was what I expected from training. The feel and the noise were well done in the simulator, and the only thing we missed was the actual physical motion during the launch. I was a little surprised at staging [first stage separation of the Saturn V] because we were told that it would be a simple matter. However, the small retro rockets fired as soon as the main engine shut down, and we went from 4.5G to -0.5G in an instant. I remember that Jim and I looked at each other in surprise, but Dave then said it was okay and that he had "forgotten to tell us about it". The rest of the launch went as predicted and we were okay when we got to orbit. I broke a little from the plan at that point, as I just had to look out the window at Earth from 144 kilometres [90 miles] up. I have to say, it was spectacular!

Being the first crew to carry the lunar rover, were there any additional concerns the crew had, for example with safety or weight, as Apollo 15 was to be the heaviest launch of a Saturn V?

Honestly, we had no concerns about carrying the rover — it was all included in the flight plan. With every mission you have to remember that you're sitting on top of what is effectively a small nuclear explosion's worth of energy [the launch energy of the Saturn V was the equivalent to the total energy used by the UK's electrical network for about 160 seconds], but we didn't see it as a problem as such — although the extra weight did force us to go into a lower orbit around Earth, as we simply did not have the energy to go any higher.

Tell us about the science experiments you conducted in orbit while David Scott and Jim Irwin were on the Moon. You said you were working 20-hour days — was it pure energy keeping you going?

I had a very extensive array of scientific experiments to conduct while in lunar orbit. They included two cameras: one a mapping camera and the other a high-resolution camera. The high-resolution camera was an obsolete camera that had been designed for the U-2 programme back in the 1950s. It took incredible photos of the surface and I was able to photograph about 25 per cent of the lunar surface. In conjunction with that camera, the mapping camera took photos of the same areas and is still being used by the cartographers to update maps of the Moon.

In addition to the cameras, I had a suite of remote sensors to scan the lunar surface. The data returned from these sensors was to allow the geologists a means of identifying the chemical content of the surface without the need to land. The rocks picked up on the surface were mostly for ground truth to match up with the remotely-sensed data. To get all this done during my time in lunar orbit I worked about 20 hours a day.

However, in free fall there is less energy involved in doing something so I did not

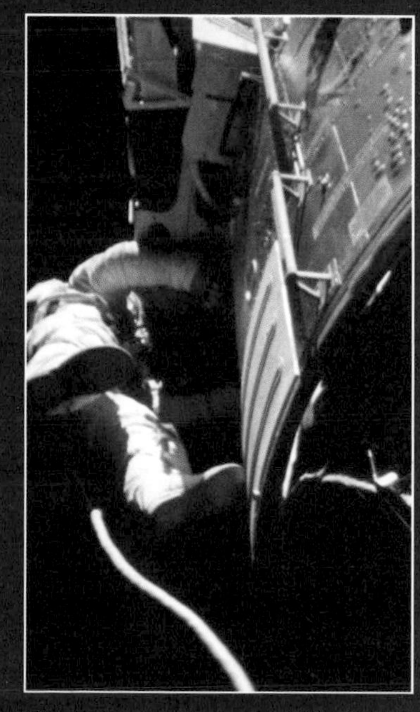
Worden floats in space outside the craft while on his 38-minute EVA during the Apollo 15 mission

Worden visits ESA's ESTEC facility and Space Expo in Noordwijk, Netherlands, on 19 October 2011

get tired. Also, the thought that we would only be there once made it important that I did everything on the list.

You hold several Guinness World Records for the 'Most isolated human being' and the 'First deep space Extravehicular Activity (EVA)'. Can you explain how it felt to be one of only three people in history to have done this and how it felt to see Earth and the Moon from such a unique vantage point? Did you get the chance for any down time while out there on your EVA?

That was a pretty unique thing to do out in deep space. But the training was so good that I felt like I was doing it in simulation. Never did I have the feeling that I was so far out. I think that is because I designed the equipment I used on the EVA, and practiced with it extensively in the zero-gravity aircraft, so I was really comfortable during the actual event. In fact, I was so well trained that it only took me about 35 minutes to complete the task.

Then I had to think of something to do so that I could stay out there for longer. I was able to get set up in foot restraints and look around, and that is when I could see both the Earth and the Moon. It was unbelievable to look at both worlds at the same time. What a unique event, and to think I was the first human being in history to see it.

Warden claims that with NASA's current plans for the SLS, a return to the Moon won't be all that productive

What do you consider to be your greatest achievement on that mission from orbit, in terms of the knowledge added to lunar geology?

There is no question about this. I had the distinct pleasure of seeing the first evidence of volcanic activity on the Moon. In my training I concentrated on looking for features that would help identify volcanic or impact features.

One of the most telling things about a volcano is that there could well be cinder cones as a product of the eruptions. I found cinder cones in the Taurus-Littrow area and not only reported them, but took high-resolution photos of them. This was probably the most important discovery from orbit. In fact, it was so important that the landing site for Apollo 17 was changed. It supported the theory that there was volcanic activity on the Moon in years past, which until then had been speculated but not proven.

Of all the flight controllers and ground crew, who gave you, as astronauts, the greatest level of support during your training and missions?

It's hard to say since it took so many different phases of the flight to train for and accomplish. They were all so dedicated and accomplished that it is difficult to pick out any one that was the most important. The simulation guys were wonderful and very knowledgeable about all the manoeuvres and software, so I'd put them very high on the list.

Meanwhile, the geologists gave us what we needed to successfully conduct the science on the flight, and the scientific investigators were superb in giving us all the information we needed to perform the experiments. I can't say any one was better than the others. I will always cherish Farouk for not only his training, but for his friendship and advice before the flight.

ESA are modifying the Automated Transfer Vehicle (ATV) to act as a modern day service module to mate with Orion, scheduled to launch in 2018, but without humans until the 2020s. Have you been inside it yet, and what are your thoughts on a return mission to the Moon?

As yet, I have not seen the ATV but I hope it is on my schedule somewhere in the future.

With the current plans from NASA with the SLS [Space Launch System], I am sorry to say, I don't see a return to the Moon as being all that productive, but I also don't know all the things that have been discovered about the Moon, so maybe there is still huge value in a return. I still believe that a large telescope on the farside would be a great thing for science, though.

© NASA; ESA; A. Doamekpor; MSFC

James Irwin's irregular heart rhythms were first noted during his Apollo mission

James Irwin

The deeply religious astronaut who found inspiration during his time on the Moon

To date, just 12 people have ever walked on the Moon. James Irwin was one of them, landing on the lunar surface alongside David Scott on 31 July 1971 in order to embark on the three-day Apollo 15 mission. It was the fourth Moon landing and Irwin was number eight to walk on the Moon, but it was the realisation of a long-held dream for the then 41-year-old pilot.

Born on 17 March 1930, Irwin had grown up wanting to go to the Moon. He studied naval science, graduating from the United States Naval Academy in 1951, and while his mother had wanted him to become a preacher, his sights were set skywards for a different reason. As such, he became an Air Force officer. This led him to aeronautical engineering and instrumentation engineering at the University of Michigan. In the late 1950s, he became a test pilot but it very nearly ended his career. In 1961 while teaching a student to fly on a training mission, his plane crashed, causing Irwin severe injuries. As well as suffering compound fractures and amnesia, he was close to losing a leg yet he recovered more than sufficiently to be selected as one of 19 astronauts by NASA in April 1966.

A couple of years later, he was assigned as crew commander of Lunar Module LTA-8. He supported the crew of Apollo 10 and he was a backup Lunar Module Pilot for Apollo 12. Apollo 15 was his crowning glory, though. It was the first mission to visit and explore the two-kilometres (1.2-miles) wide Hadley Rille canyon and the 4.5-kilometres (2.7-miles) tall Apennine Mountains. The Lunar Module was on the surface for a then-record 66 hours and 54 minutes.

Irwin and Scott collected an astonishing 77 kilograms (170 pounds) of lunar rock samples as they carried out inspections of the nature and origin of the area, spending 18 hours and 35 minutes on the surface during the course of three extravehicular activities. Since the first lunar rover accompanied them, they were able to travel further from the Lunar Module than previous missions could and their endeavours led to the discovery of the Genesis Rock, which was found to have formed in the early stages of the Solar System.

Deeply religious, Irwin found his time on the Moon to be very moving. After the first day of exploration, he said the landscape reminded him of his favourite Biblical passage from Psalms. In reciting this — "I'll look unto the hills from whence commeth my help" — he then displayed a good sense of humour, adding: "But, of course, we get quite a bit from Houston, too." Apollo 15 successfully splashed down in the Pacific Ocean, with Irwin having logged 295 hours and 11 minutes in space.

Apollo 15 was viewed as a very successful mission. It was the first to use a lunar surface navigation device, the first to see a sub-satellite launched in lunar orbit, and it saw the first scientific instrument module bay to be flown to the Moon. Yet it was also Irwin's first and only time in space. He resigned from NASA and the Air Force in July 1972 and he went on to form a religious organisation called High Flight Foundation in Colorado Springs.

Unfortunately, the following year he was struck by ill health, suffering a heart attack while playing handball. He had another cardiac irregularity in 1986 while running and then fell prey to a fatal heart attack while he was riding his bike on 8 August 1991. He never recovered and he died the same day, becoming the first of the dozen men to walk on the Moon to pass away.

Duke became the youngest person to walk on the Moon at the age of 36

Youngest man on the Moon

On 16 April 1972, Apollo 16 left Earth and headed to the Moon. The mission's Lunar Module pilot Charles Duke recounts the day his astronaut boot touched the lunar soil

Interviewed by Gemma Lavender

Could you tell us a bit more about how you went about choosing the landing site without crashing?

There is no dark side of the Moon, it's actually the far side. Obviously it rotates once every 28 days. There are two weeks of daylight and two weeks of night on every spot of the lunar surface. Apollo 16 landed with a low Sun angle to give us [the Apollo 16 team consisted of Duke, commander John Young and Command Module pilot Ken Mattingly] definition of the lunar surface. If you tried to land at high noon, it was all washed out [by sunlight], which meant that you couldn't see any of the craters and you couldn't see any of the elevation changes. The landing site was therefore chosen at a very low Sun angle, so that we had all of the shadows to the West. It was early morning during the Moon day at the Apollo landing site, which was called Descartes. We got some definition of the landing site, which meant that we didn't crash or fall into a big crater.

The further east you go, the more the backside of the Moon was in darkness. We landed just a little east and a little south of the centre and could see that half the backside of the Moon was in sunlight.

What did you see as you entered lunar orbit?

As we entered the shadowed portion of the Moon, you got this eerie feeling because the Sun hadn't been shining on this region for a few days. The feeling was so unreal that I was left thinking: 'Well, I hope our tracking is right!'

You're going into orbit at 60 by 70 miles [97 by 113 kilometres] above the Moon, and so we burned to slow down and manoeuvre into orbit. At this point, the computer told us that we were out of contact with the Earth and that we had loss of signal. Then, all of a sudden there was the sunrise — it was the most dramatic sunrise I've ever seen. In Earth orbit, you see the Sun's glow on the horizon or the planet's atmosphere and it gets brighter and brighter. The Moon is different, though; there's instant sunlight with long shadows on the lunar surface. The far side of the Moon was very rough back there — I would not have wanted to land on the backside of the Moon.

The manoeuvre into orbit [when we were close enough to the Moon] lasted around two minutes and 41 seconds. During that time, we burned about 2 million kilograms [4.4 million pounds] of fuel. We took pictures of the Earth [from lunar orbit] after we left our planet over Australia around an hour after lift-off. Earth was like a jewel suspended in the blackness of space. The Sun shines all of the time on the way to the Moon but the stars are never visible. It's very dark when you look outside and all you see is the Earth, Moon and Sun.

Why didn't you land on the far side of the Moon?

We wanted to be in contact with the Earth, so we weren't able to land on the backside of the Moon. We ended up landing at a place called Descartes — our landing site was in the Descartes Highlands. We were the fifth mission to land on the Moon and I can say that it really is a dramatic place.

We had to lower the spacecraft further down into orbit by about 60 by eight miles [97 by 13 kilometres]. That would be the orbit in which we would attempt to land. At this point we needed a critical burn of the main engine because if you had one second overburn, you ended up impacting the Moon's surface.

Interview Bio

Charles Duke

As Lunar Module pilot for Apollo 16, Charles Duke became the tenth and youngest person to walk on the Moon at the age of 36, when he landed on its surface in 1972. An engineer, retired Air Force Officer and test pilot, Duke has spent over 260 hours in space.

The Apollo 16 Saturn V vehicle carrying the mission's astronauts

Duke is pictured here collecting lunar samples at the Descartes landing site

"If the roving vehicle broke down then we would have had to walk back - there's no rescue on the Moon!"

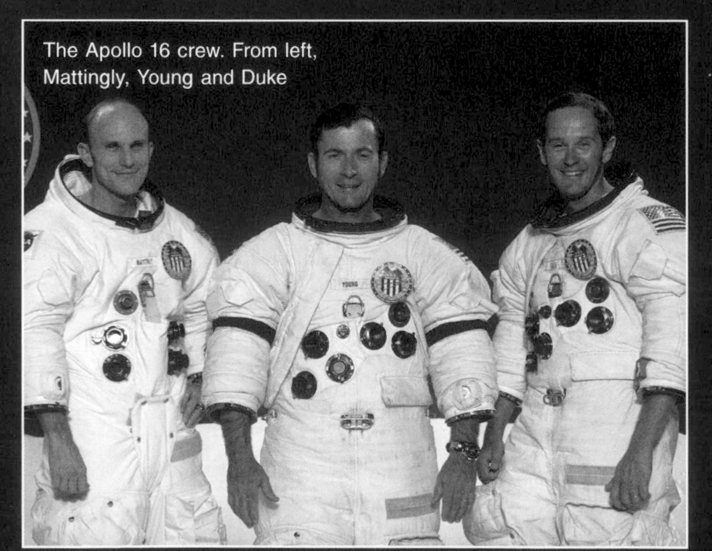

The Apollo 16 crew. From left, Mattingly, Young and Duke

As we were lowering ourselves, we looked out of the window as we came around from the backside of the Moon to make sure that we weren't going to crash into its surface. If we felt like we were going to, then we would have had to bail out.

Your Command Module pilot Ken Mattingly was meant to go on a mission to the Moon before you. What happened?

[Laughs] Mattingly was supposed to be on Apollo 13, but I caught measles a week before the launch and exposed him to it, so they jerked him off the flight and we ended up on a flight together.

What was lift-off like?

Everybody went to the Moon on a Saturn V rocket. At lift-off the engines were producing 3.5 million kilograms [7.7 million pounds] of thrust, so you didn't lift off very fast but you were shaking like crazy from side to side. I got a little bit nervous — you don't see outside of Apollo at this point. The windows were covered over, so I got a little bit nervous and my heart was pounding. I said "I hope this thing makes it", because the vibration was just so intense.

Later on, when we returned to Earth, I spoke to the flight surgeon who was based at mission control and asked what my heart rate was. He said: "It was 144 beats per minute — you were excited!". I replied: "You bet I was excited!" I then asked what John [Young]'s was and, apparently, his was 70. So he was the cool one.

What did you do as soon as you landed?

We were supposed to go outside right after we landed, but we didn't get the chance. Instead, we went to sleep and had a rest period. That was pretty hard to do, three hours after we landed on the Moon. With the help of a sleeping pill, though, I was able to do it.

Could you describe the moment you left the Lunar Module?

I opened the door, got on my hands and knees, and crawled out backwards down the ladder and onto the footpath. From the door to the footpath, to give you some idea, is about five metres [16 feet] in height. Descartes had some mountains and plains as well as a big valley, which was about ten kilometres [6.2 miles] wide. We would explore this valley over our three-day stay. You could see the shadows of the Sun, in the direction of Stone Mountain. The Sun was always in the west while we were on the Moon's surface.

We drove further afield on our second EVA [extra-vehicular activity]. It was about four kilometres [2.5 miles] from where we landed. We were actually going to go out as far as 100 kilometres [62 miles], but we didn't drive that far because if the roving vehicle broke down then we would have had to walk back — there was no rescue up on the Moon!

How did the Moon feel under foot?

The Moon is covered with this very fine dust, like a powder, which is actually pulverised rock. When we walked on it, we didn't sink any further than a couple of inches. One of the problems in the beginning was that some of the scientists thought that we were going to land in lunar dust that was 200 to 300 metres [656 to 984 feet] deep and we were going to sink when we stepped outside.

Did you get a chance to study the Moon's surface?

We could see the western horizon from our location. And we could see a place called North Ray crater, and all of the white rocks in our area were from the meteor impact that made that crater. We collected three colours of rocks — grey, white and black. Lesson number one in geology is to pick up a rock in every colour and so we were able to get the whole suite of Moon rocks. We collected about 98 kilograms [216 pounds] of lunar soil. We also had a whole suite of experiments. Two seismic experiments [to measure seismic waves through the Moon], a

Duke as Apollo 11 CAPCOM in 1969

mass spectrometer [to figure out what the lunar soil and rocks are made of] and a heat flow experiment, which unfortunately failed. We also had two magnetometers [to measure the magnetic field].

To the west of where we landed was Plum Crater. It was about two metres [6.6 feet] deep — or perhaps a bit deeper. We gave it a wide berth because we didn't want to fall into it since, as I said earlier, there is no rescue on the Moon and we wouldn't be able to get out. We got a good view of the lunar surface during the mission, we got plenty of samples and a good view of the lunar colours — there were different shades of grey. We collected rocks with different shovels, rakes and tongs.

We could see for a long way from Stone Mountain, which is the furthest we got on the mission, and the lunar plains glistened in the sunlight. It was very bright on the Moon, so we had our visors down at all times. In some places, our footprints were barely there because the lunar soil was so thin. I drilled three holes in the Moon — around three metres [9.8 feet] deep — and the soil was still quite solid. In some areas on the lunar surface, it was like walking on a solid floor. There was no dust at all.

What did the Earth look like from the Moon?

From where we stood on the Moon, the Earth was directly overhead. And it stays there. I did see the Earth from lunar orbit, but not from the Moon. If you look up, you're looking at the top of your helmet — this part of your spacesuit doesn't move back with your head. I only ever saw the Earth when I fell backwards one time — it scared me half to death! I fell back, and said: "Oh, yeah, there it is!" It's a good job that I knew how to get back up again because I would still be on the Moon.

Apollo 11's Neil Armstrong and Buzz Aldrin left a flag on the Moon. Did you leave your mark in a similar way?

I left a picture of my family on the Moon. People wondered why I did that but I really wanted to get my family involved in the mission. We were travelling and training all of the time [so I rarely got to see my family] and so I asked them: "Would you like to be on the Moon with me?" My children said: "Yes, sir, we sure would!"

So we took this picture of the family and on the back of the picture, we wrote: "This is the family of astronaut Duke from Planet Earth. Landed on the Moon, April 1972." We all signed the picture. It's still there. The lunar rover is still there also and we left the TV camera running on the front of it. The rover could travel at a maximum speed of 17 kilometres per hour [10.6 miles per hour]. Up on the Moon, which only has 17 per cent of the Earth's gravity, it felt like you were flying.

On your return to Earth, where did you land?

We splashed down in the Pacific Ocean and the three parachutes started to open before fully deploying as we landed in the water. It was one of the most beautiful sights I ever saw [laughs]. We were grateful for the parachutes — without them, the spacecraft would have hit the water with such force that Apollo 16 would have split open and we would have sunk to the bottom of the ocean.

Duke left a picture of his family on the lunar surface

A view of the Earth from Apollo 16

Commander John Young (pictured here with the lunar rover) was also a crew member of Apollo 16

© NASA; ESA; Flickr

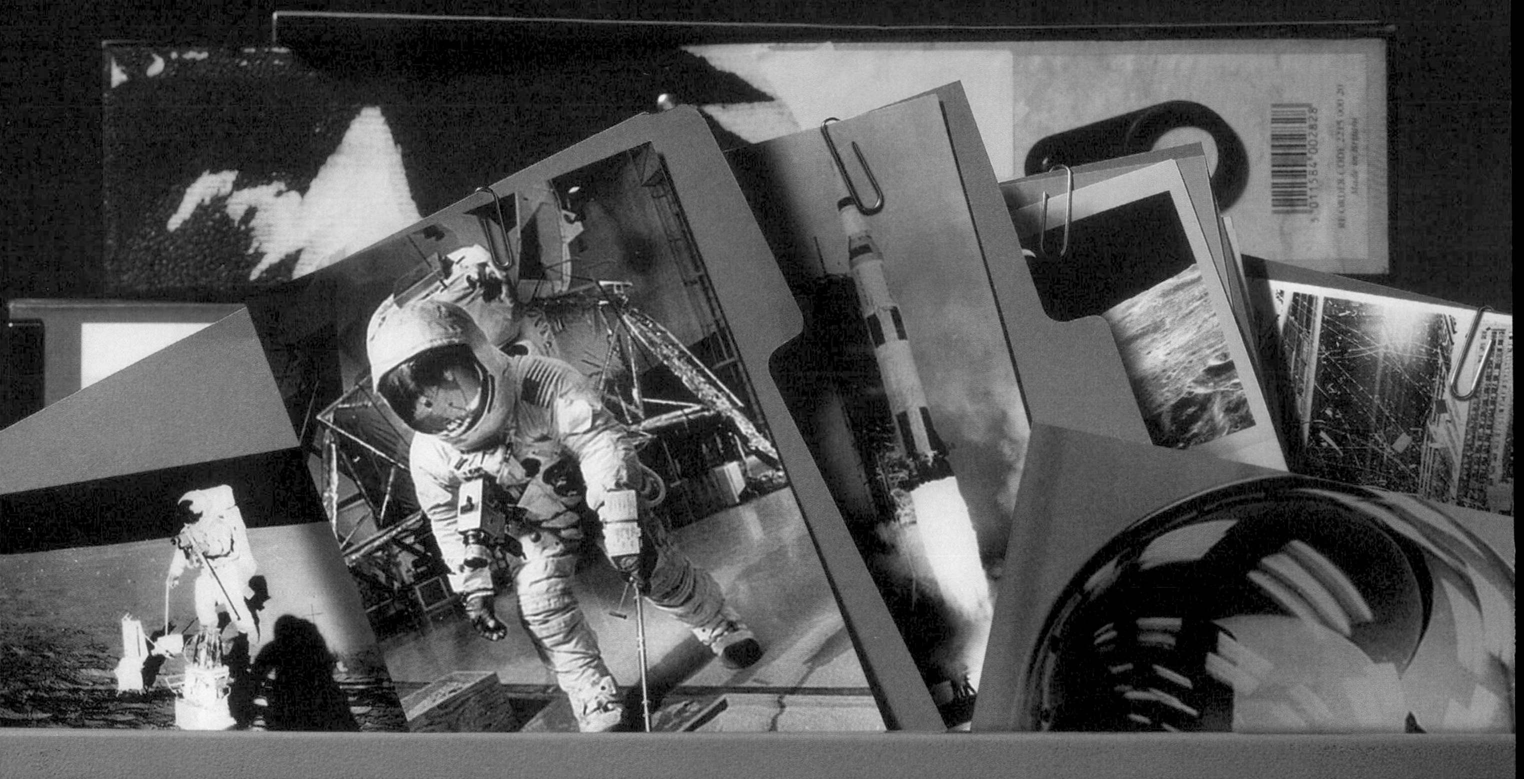

THE HIDDEN ARCHIVES OF APOLLO

A selection of images from the Apollo era have risen from the NASA files - and they're here to see, courtesy of Swann Auction Galleries

The Apollo archives are teeming with images that mark what is considered to be the golden age of space exploration — the first manned lunar exploration missions in the history of humanity. But wait… there's more.

NASA and the Swann Auction Galleries have recently released a never before seen collection of images taken from the Apollo Program. Between the years of 1969 and 1972, humanity explored a new frontier over the course of six manned spaceflights to the Moon, resulting in 12 American astronauts walking on the Moon, but unfortunately we haven't explored it since. This makes for a rare photo album to feast your eyes upon sights not many have seen before.

THE BEGINNING OF A NEW ERA

The return of Apollo 11's Lunar Module signified the end of the beginning, as this event brought a whole new dimension to the space age.

Leaving the Moon

After making history, this photo was taken from the Command Module by its pilot, Michael Collins. The half-shining Earth watched over the Lunar Module containing Neil Armstrong and Buzz Aldrin as it slowly made its return to the Command Module.

The first two moonwalkers had just spent over 21 hours on the lunar surface, of which roughly 2.5 hours were part of the moonwalk itself. The pair began their return to the Command Module at 17:54 UTC on 21 July 1969. A few hours later, Collins captured this inimitable image that unites the Earth, the Moon and humankind.

> **"This mission continued the exploration of the lunar surface that its predecessors, Armstrong and Aldrin, had started"**

Lunar landing simulation

On 27 October 1969, NASA photographed the eventual Apollo 12 commander, Pete Conrad, undergoing intense training designed to simulate a lunar landing. This mission continued the exploration of the lunar surface that its predecessors, Armstrong and Aldrin, had started. However, in order to do so, Conrad had to make sure they could firstly get there safely.

The Lunar Landing Research Facility (LLRF) at NASA's Langley Research Center played a crucial role in preparing the Apollo astronauts for the flight, landing and walking on the Moon's alien environment. The surface even had craters on it in order to make the simulation more realistic.

© NASA

Apollo 12 men at work

While busy at work on the Moon, Pete Conrad took this great shot of Alan Bean. Here Bean is holding the vacuum-sealed Special Environmental Sample Container (SESC) that has been filled with lunar soil for study back on Earth. This picture was taken in November 1969 at Sharp Crater.

From the angle of this image you get a view of Bean's Hasselblad camera, which has been mounted onto the chest of his spacesuit. Conrad's reflection can also be seen in Bean's helmet visor, as well as the Hand Tool Carrier.

Apollo 13 liftoff

The third lunar landing attempt was to be a "successful failure". The Apollo 13 crew consisted of James Lovell, Fred Haise and John Swigert, who intended to land in the Fra Mauro region of the Moon. But, two days after launch an oxygen tank exploded and they were forced to abandon the landing. Instead they orbited the Moon, and eventually returned safely to Earth.

On 11 April 1970, none of the astronauts aboard the Saturn V rocket at Kennedy Space Center in Florida, United States, thought they'd have such an ordeal ahead of them.

Apollo 12 success

This photo was taken when Pete Conrad and Alan Bean walked on the Moon, making them the third and fourth men respectively to do so. These two astronauts doubled the number of Extra Vehicular Activities (EVAs) performed by Apollo 11... to two. Each moonwalk lasted roughly four hours.

On the first EVA, Conrad and Bean deployed part of the Apollo Lunar Scientific Experiments Package (ALSEP), and this is what is photographed here. ALSEP was placed at each Apollo landing site — except Apollo 11 — with the aim of relaying long-term data from the lunar surface.

WHAT ABOUT A PHOTO?

It is a little-known fact that Bean and Conrad also had a colour video camera on the Moon, but it was broken after Bean pointed the camera at the Sun, ruining the optics.

Crescent Earth

Shown in this particular image is the crescent Earth shining over the uneven cratered surface of the Moon. It cannot be said with certainty what mission this is due to the lack of image description, however, based on the images of other Apollo missions, this one was most likely taken on the last ever mission, Apollo 17.

This conclusion is drawn from Apollo 17 images that were taken from the Apollo 17 Command Module while in lunar orbit. If true, this image would be the most recent image out of the collection.

Celebrations for Apollo 11's astronauts

(From left to right) Aldrin, Armstrong and Collins returned to Earth as heroes. These were the first men to not only travel to the Moon, but also have two of the three descend to its surface and walk where no one had ever walked before.

These three brave men were welcomed back – after 21 days in quarantine – with a ticker tape parade through the streets of Chicago, United States, where an estimated 1 million people turned up on the 13 August 1969. Here, the trio are pictured travelling down LaSalle Street, Chicago.

Over the Moon

Another Apollo mission, another launch from the Kennedy Space Center. At the tip of this particular Saturn V rocket would eventually sit Apollo 12's commander Pete Conrad, Lunar Module pilot Alan Bean and Command Module pilot Richard Gordon. At the time of the image, the Saturn V sat empty.

Overlooking the Apollo 12 rocket here was their target, roughly 380,000 kilometres (236,000 miles) away. The Saturn V rocket, shone brightly upon by searchlights, was pictured during a test countdown on the 27 October 1969, which had the service tower move away during the process of the test.

© NASA

"In this image you see the mystery astronaut simulating lunar soil collection, along with the Lunar Module in the background"

Preparing for lunar science

There are images in this selection that show astronauts actually on the Moon, collecting samples and conducting real science, although this is the only image that shows them practising these procedures.

Due to the lack of image description, it cannot be said with certainty who this astronaut is, or what mission this is. After going through the other archive images, the best bet is either Jim Lovell or Fred Haise preparing for Apollo 13. In this image you see the mystery astronaut simulating lunar soil collection, along with the Lunar Module in the background.

James 'Jim' Arthur Lovell Jr

NASA pictured Jim Lovell, flight commander of the Apollo 13 mission, prior to his departure from the Kennedy Space Center. Suited up in the iconic Apollo spacesuit, Lovell would not actually land on the Moon, but along with his crew would navigate the damaged Command and Service Module, Odyssey, back to Earth.

Approximately 330,000 kilometres (205,000 miles) from Earth, the crew were forced to abandon Odyssey and use Aquarius — the lunar module — as a 'lifeboat'. But the question was whether these astronauts would come home alive? Six days after they left Earth they did, in what has been termed NASA's finest hour.

THE APOLLO SPACESUITS

The spacesuits that were worn on each Apollo space flight were made up of different layers of nylon, aluminised Mylar, Kapton and Teflon for the ultimate protection.

Looking back at Apollo 11

At the landing site referred to as 'Tranquillity Base' on the Moon, Neil Armstrong snapped this picture of his fellow astronaut Buzz Aldrin, as indicated by the name on his chest piece. When looking at his reflective faceplate, a lot more of the scene is unveiled, including Armstrong himself, a section of the Lunar Module, the solar wind experiment and the United States' flag. The fact that this picture is from one of — if not the — most iconic events ever, and hasn't been released until recently makes it all the more surprising and extraordinary.

© NASA

Interview Bio

Dr David Parker

Dr David Parker is the former chief executive of the UK Space Agency and ESA's current director of human spaceflight and robotic exploration. He has a PhD in aeronautics and astronautics from Southampton University and began working for British Aerospace Space Systems in 1990 on technology research for missions. During his career, he has worked for Matra Marconi Space in Bristol and he was assistant director at the British National Space Centre.

Man's Return to the Moon

NASA plans to take astronauts around the Moon for the first time since 1972 with the European Space Agency. ESA's director of human spaceflight and robotic exploration, Dr David Parker, speaking in 2017, told us more

Interviewed by David Crookes

ESA is working on Orion — a spacecraft NASA intends to use to send astronauts into space in 2021. It is building a cylindrical unpressurised service module, which will provide electricity, water, oxygen and nitrogen while keeping the craft on course, and it is due to embark on a test mission next year. But how did ESA get involved?

ESA is involved in the International Space Station (ISS) — we have astronauts up there and we carry out science and technology on board. We are also eight per cent contributors to the ISS, but there are certain things that we don't do at the moment, such as launching astronauts and shipping cargo to and from the space station. It means that we have to barter in order to use these kinds of facilities and tools with our NASA colleagues.

In this case, we were looking long term towards exploration beyond low-Earth orbit and the idea came about of participating in NASA's Orion programme. We felt we could build on the know-how we'd built up in developing our Automated Transfer Vehicle, which serviced the ISS between 2009 and 2015, and that we could bring our knowledge to the table. Basically, in exchange for European industry building the service modules for what will be Exploration Mission 1 (E-M1) and Exploration Mission 2 (E-M2), we get to continue to work on board the ISS and fly astronauts there. It fulfils our access to the ISS and it gets us involved in the future of exploration.

> "It was very much the long-term ambition of ESA to have astronauts heading out beyond low-Earth orbit"

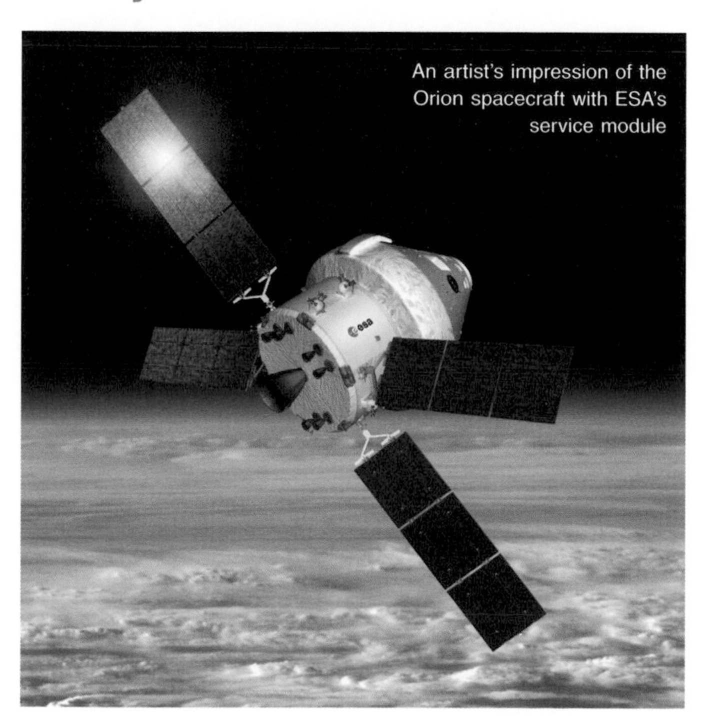

An artist's impression of the Orion spacecraft with ESA's service module

It is going to be the first deep-space exploration since Apollo in 1972. How does that feel?

It's super exciting for exploration in general and it's super exciting for Europe because it will literally be the first vehicle built by Europe that carries astronauts. All of the skills and knowledge needed to meet the requirements of carrying astronauts also brings its own challenges but we've just shipped the propulsion demonstration module and we're on the critical path of NASA's exploration programme, which is really motivating for the industrial teams. It's also a big vehicle. The whole fuelled mass is 35 tons.

Has going beyond low-Earth orbit been ESA's ambition for a while?

Yes, it has been an ambition for a long time to be involved in deep-space exploration. There are [ESA] studies going back ten years on lunar exploration and of where we go next, so it was very much the long-term ambition of ESA to have astronauts heading out beyond low-Earth orbit. Of course, we still need to stay in low-Earth orbit to do the science and research that happens there. But increasingly that may move to a more commercialised model. We're already seeing [in the US] commercial companies involved in taking cargo and, eventually, crewed vehicles to the ISS. But the vision is to go further, definitely.

The first astronauts leaving low-Earth orbit will be from NASA. Will ESA astronauts be looking to go?

As far as the agreement goes today, the first crewed mission will involve NASA astronauts — we are building something and they will get to fly it. But eventually one day the ISS will come to an end and the sort of ideas being talked about are concepts of a deep-space habitat — the idea that we have a crewed vehicle in orbit that is able to carry four people with at least a laboratory and an electric propulsion system to be in orbit and do shakedown cruises in deep space. Orion would carry the astronauts to that deep-space habitat and dock with it. So you start to build up this idea and it's certainly an ambition. Nothing has been agreed but we want to be part of that.

The service module sits directly below Orion's crew capsule. It is 5m (16ft) wide and 4m (13ft) high, weighing 13.5 tons without propellant

"Part of the exploration is to do science and technology innovation and inspire people —international cooperation is at the centre"

The Orion spacecraft lifts off on a test launch in December 2014. It does not have the European Service Module attached

Is cooperation between ESA and NASA strong?

Part of the exploration is to do great science and technology innovation and inspire people, but international cooperation is at the centre. Everything we do at ESA is in international cooperation, whether we are working with NASA, Russia or exploration partners in the station. It's essential to share the burden and carry the workload. Working alone, we could never contemplate building something like Orion so to be part of it is a real bonus.

Tell us more about the hopes of 2018's E-M1.

It's going to be the testing of NASA's Space Launch System (SLS) so there's going to be an enormous rocket — the biggest thing since the days of Saturn V. Simply seeing that launch will be extraordinary, but it will test the behaviour of the Orion module, its loop around the Moon and its return. It will also test the behaviour of the service module and its propulsion system, which has a large engine that is derived from the Space Shuttle's Orbital Manoeuvring System engines. So it's testing the end-to-end system, and obviously the tracking systems and the control systems. E-M2 will have the added factor of astronauts aboard. In that instance, the astronauts will take the first views of the Moon from the digital age back to Earth. It will be a real milestone.

What is expected during E-M2?

The current plan is an eight-day mission. It goes up into a parking orbit around the Earth for a bit over a day so there's a three-and-a-half-day outbound to the Moon and then coming back again. So it adds up to about an eight-day mission at least with a free return so that if there are any problems, there are no manoeuvres by the vehicle applied at the Moon. It will probably have some piggyback payload, such as small CubeSats put on a trajectory to the Moon. Some of them will try and go into lunar orbit and there may be some other larger payloads on board as well.

Has there been a lot of frustration that humans haven't been out of low-Earth orbit since 1972?

I think if you talk to Buzz Aldrin and anyone down from that, they'll say we're waiting for that experience [to be replicated] in the digital age. I remember the Apollo missions but we were watching it on grainy black and white televisions. This will be the digital age where everybody globally will be able to watch it in real time. There will be that sense of looking up at the Moon and knowing that humans are travelling there and it will be an amazing experience.

What are the plans after this?

In terms of human exploration, the end is sending people to Mars. NASA talks about a journey to Mars but "journey" is the important word. There's a lot to learn before we can make that voyage. If you think about it, the ISS has astronauts in a comfortable working environment — there's a large volume and a regular supply of food, oxygen and other supplies. But when you think about living and working in deep space, you have to consider how you can reduce the amount of logistic supplies and about better efficiencies in recycling water and waste, and in generating oxygen. We also have to tackle the challenge of radiation protection. The Apollo astronauts went to the Moon and back within a few days and were lucky enough not to be exposed to high doses of radiation. But we'll have the challenges of radiation protection on longer missions.

Will there also be issues of mental wellbeing?

Yes, of course. Rather than the wide-open space of the ISS, astronauts will be constrained in a small volume. It'll be more like a deep space campervan than a luxury hotel space station. So we have to work out how to support the astronauts, transport them into deep space, protect them from the radiation, supply them and keep them mentally and physically in good shape. It'll be a step-by-step vision that reaches out with robots and humans working together, going beyond where we are today.

What are your hopes for the ISS though?

Europe has decided to carry on being a major part of the ISS until 2024 and there is a lot of very good science that's planned now, whether it's biology related or developing materials. We're starting to see industrial companies looking at the feasibility of manufacturing in low-Earth orbit, such as high quality optical fibres that can be potentially manufactured in a space environment. Looking beyond that, there will be a gradually greater involvement of the commercial sector in using low-Earth orbit. You can envisage a whole world of commercial Space Stations, of commercial crew and cargo vehicles and uncrewed vehicles doing science in space. I think what we will see is a diversity of things happening in low-Earth orbit rather than one big piece of infrastructure.

What frontiers are you pushing technologically?

In terms of low-Earth orbit, how we get more efficient in terms of doing the science and recoverable vehicles is an area of interest — the potential for launching and recovering vehicles and reusing them is important. For going into deep space, we're interested in the habitat modules and very high-powered electric propulsion. We're doing a lot of electric propulsion today for the scientific missions at 10kW and 20kW levels. We need to look at 30kW and 40kW size engines for deep-space exploration. Then there's the interaction between humans and robots. One of the experiments that [ESA astronaut] Tim Peake did last year was demonstrating the control of a planetary rover on the surface of the Earth but in a simulated environment and controlling that from the ISS. So that is kind of projecting forward.

You mentioned Tim Peake. He really caught imaginations with the work he did with schools and his interactions with the public. How important has his contribution been?

When I was running the UK Space Agency, working with schools was a very big part of our ambition. Tim Peake did a fantastic programme of science but we really wanted a legacy for the next generation, so a big effort was made with a lot of different partners to deliver a fantastic education programme that appealed to different ages of school students.

It wasn't just science and technology but things that involved arts and humanities and it had a massive impact. The UK Space Agency estimates that 1 million kids were involved in activities around his mission. From the moment of launch to the end, it attracted a lot of attention for the next generation because the kids in the UK hadn't experienced anything like this. I go back to my own childhood and remembering Apollo, and this was their Apollo, I think.

A test version of ESA's service module is put through its paces in Ohio, US. It will sit on top of the Space Launch System and contain over 2,500 tons of propellant

This is the real European Space Module being created in the assembly hall at Airbus Defence and Space, in Bremen, Germany

There has been a real resurgence in space. Why?

I think it's something to do with the diversity we have now. It's not just NASA. There's so much happening in Europe, India, China and Russia and you have all of the commercial billionaires like [Space X's] Elon Musk and [Amazon boss] Jeff Bezos developing their own rockets.

A new generation is seeing the excitement and limitless potential of space and how it relates to us on Earth, and it appeals to the heart as well as the brain. It feels like it's a very dynamic time again and things are happening very quickly after a long period where it appeared to move very slowly.

What's happened since 2017?

While the Orion mission is still going ahead, it has been pushed back to 2022 — but something arguably more exciting will be taking place, too. In March 2019, NASA announced its intention to put man back on the Moon by 2024. Citing the lunar South Pole as the destination, NASA adminstrator Jim Bridenstine has announced that "I have already directed a new alignment within NASA to ensure we effectively support this effort, which includes establishing a new mission directorate to focus on the formulation and execution of exploration development activities. We are calling it the Moon to Mars Mission Directorate." The plan is to head further into space in the years to come, but first, NASA is returning the the Moon, "this time to stay".

© Photoshot; Gary Lee; ESA; D. Ducros; J. Harrod; Airbus DS; NASA; B. Ingalls

Living on the Moon

How we could turn craters into colonies for human life

The Moon is our closest neighbour, but only 12 people have ever set foot on its surface. Since 1972, the only visitors have been robots, orbiters and probes. For a long time there was little interest in going back, but at just three days journey away from Earth, the Moon is an obvious target for further investigation. With more countries establishing their own space programmes, and an increasing number of private companies entering the field, interest in the Moon is growing once again.

The environment on the Moon's surface is hazardous, but if we can find a way to construct a base we would gain access to a wealth of off-world resources. It is a prime location for telescopes and communications equipment, and its unique environment could hold clues to the history of the Solar System. The Moon's potential has been recognised by organisations across the world, and there are now several exploratory missions in development. At the moment, these are focused around finding out more about the Moon's potential, but over the next few decades, manned missions and even base construction could be on the agenda. Russia's Roscosmos are planning a series of Luna-Glob missions as a starting point for establishing a robotic base, and in collaboration with the European Space Agency, they are hoping to scope out the Moon's south pole in 2019 and 2020. The China National Space Administration are developing a series of Chang'e probes to collect lunar samples in preparation for future mining missions, and they are building a shuttle capable of lifting human astronauts to the Moon. What's more, in 2007, Google launched the Lunar XPRIZE, encouraging private companies to land rovers on the surface by 2017.

Even NASA, who has chosen to focus their resources on manned missions to asteroids and to Mars, are developing a probe to map the water deposits on the lunar south pole. At the moment, we are just taking our first tentative steps towards further exploration of the Moon, but in the future a science fiction-style base on the surface could become a reality. We explore what such a lunar outpost might look like, and what hazards and challenges could get in the way.

Why the Moon?

With preparations already underway for manned missions to Mars, some might question the logic behind a return to the Moon, but a lunar outpost could bring several advantages. A trip to the Moon and back could be completed in under a week, and the surface is rich in resources. Lunar dust contains hydrogen, oxygen, iron and other metals, and if these resources could be mined, it could provide a close off-world source of water and building materials.

The far side of the Moon is shielded from the noise of Earth's communications, providing a quiet vantage point for looking out into the universe, and the near side has a constant view of the surface of our planet, making it an ideal place to set up monitoring stations. Navigational support could also be provided for a variety of operations, from search and rescue on Earth to deep space exploration. A base on the Moon would also allow us to look closer at its geology, which in turn would help us uncover more about its history and the evolution of the Solar System. Experiments could be conducted, and materials and equipment could be tested, away from the familiar conditions on Earth.

Lunar holidays

With space tourism barely in its infancy, it might seem a bit premature to consider the idea of holidaying on the Moon, but if humanity were to establish a base up there, visitors would almost be inevitable. The company Space Adventures has already sold two $150 million tickets for a trip to visit the Moon, and more private organisations are looking to set up their own tours.

Rules set out in the 1967 Outer Space Treaty state that the Moon can't be claimed by any country, even if they have set up a base there. However, laws regarding the exploitation of the Moon and its resources for commercial gain have not yet been fully established.

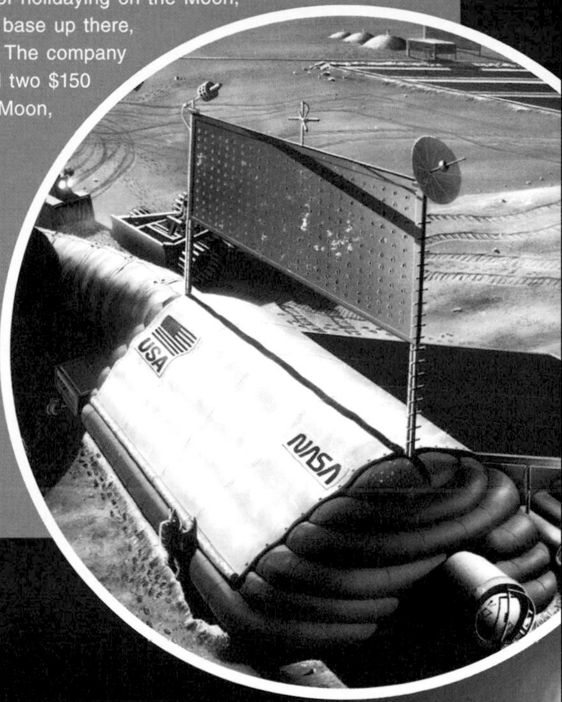

A base on the Moon could pave the way for a new kind of holiday

Colonising space

A lunar base could perform many different functions, from mining to communications

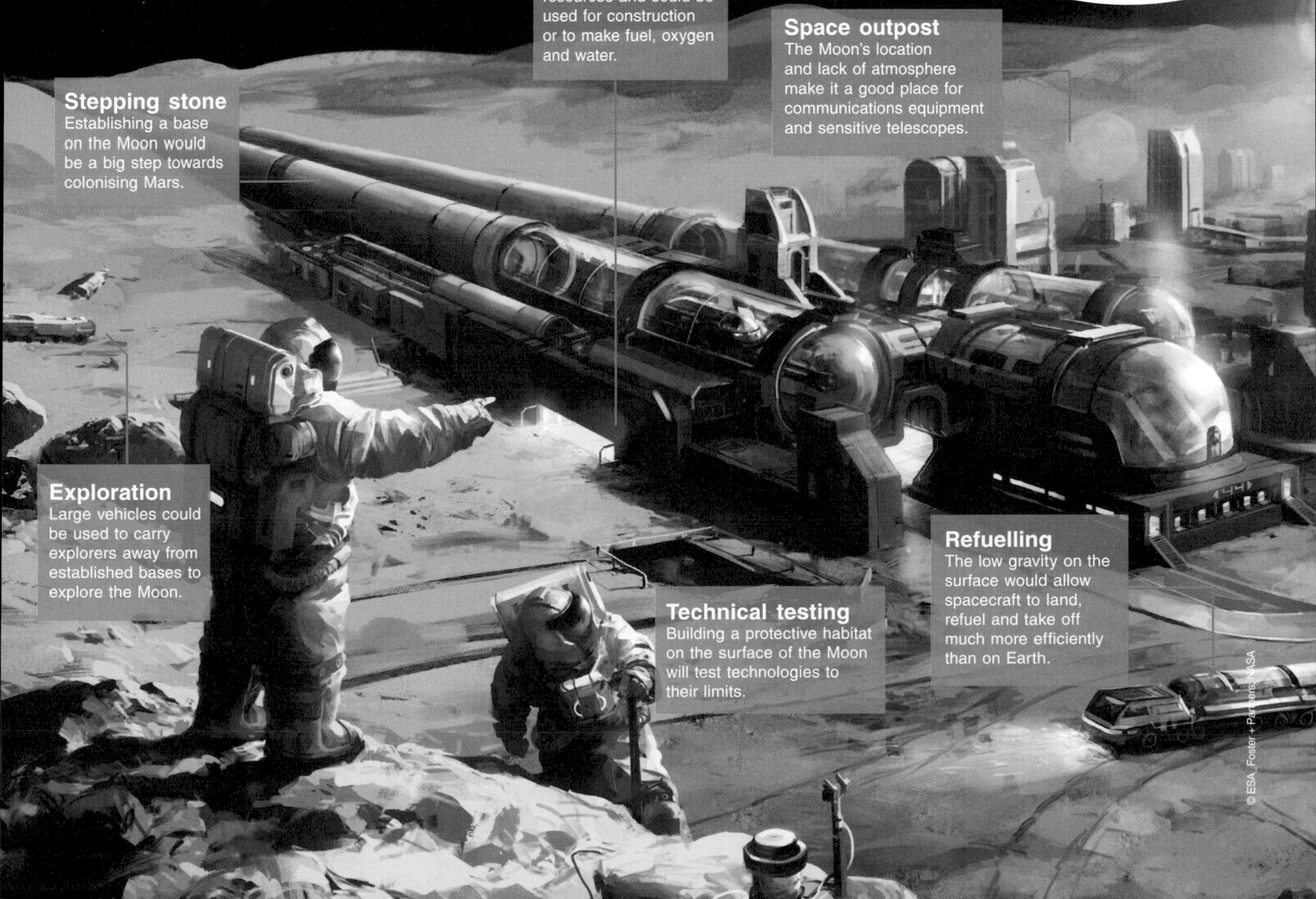

Mining and excavation
The Moon is rich in resources and could be used for construction or to make fuel, oxygen and water.

Space outpost
The Moon's location and lack of atmosphere make it a good place for communications equipment and sensitive telescopes.

Stepping stone
Establishing a base on the Moon would be a big step towards colonising Mars.

Exploration
Large vehicles could be used to carry explorers away from established bases to explore the Moon.

Refuelling
The low gravity on the surface would allow spacecraft to land, refuel and take off much more efficiently than on Earth.

Technical testing
Building a protective habitat on the surface of the Moon will test technologies to their limits.

© ESA_Foster + Partners; NASA

How to build a base

The Moon has little atmosphere and none of the protective shielding that we enjoy here on Earth; as a result, the surface is hostile. It is pummelled by solar winds, scorched by radiation, and chunks of rock regularly fall from the sky. The ground is coated in the shattered remains of ancient asteroid impacts, forming a thick layer of sticky dust, and with no atmosphere or weather to wear the particles down, the grains are razor sharp. A successful base would need protection against all of these threats, and, for people to stay there long-term, it would also require a steady supply of food, water, oxygen, power, shelter and rocket fuel.

One of the most popular concepts for a lunar base is inflatable housing — lightweight and easily assembled by pressurising from the inside. With the airlock from the landing capsule used as a door, these structures could provide a quick and simple solution to setting up a base. However, a puncture could prove catastrophic, so the pods would need to be shielded in underground chambers or beneath piles of Moon dust. Flat-packed panels could also be shipped in from Earth to build sturdier dome or hangar structures, but it would be much more fuel-efficient to use building materials found on the surface of the Moon. When heated, lunar dust can be transformed into a tough solid that could be used to construct buildings and roads, and 3D printers could one day be used to make structures from the regolith. In the right location, solar panels could provide renewable power for the base, and, if plants are able to grow on the Moon, it could one day be possible to set up a semi-sustainable farming and composting system. Then, if water, oxygen and hydrogen (rocket fuel) could be extracted from lunar dust, a base might even be able to become self-sufficient.

Unfortunately, there are still major challenges to be overcome before we reach this stage, not least the devastating effects of lunar dust. The dust seems to find its way inside even tightly sealed spaces, causing rapid damage to equipment. There are some ideas to get around this, including cable cars or covered transport tubes to minimise the disturbance on the surface, and clean rooms and air locks to keep inside spaces dust-free.

"Solar panels could provide renewable power for the base"

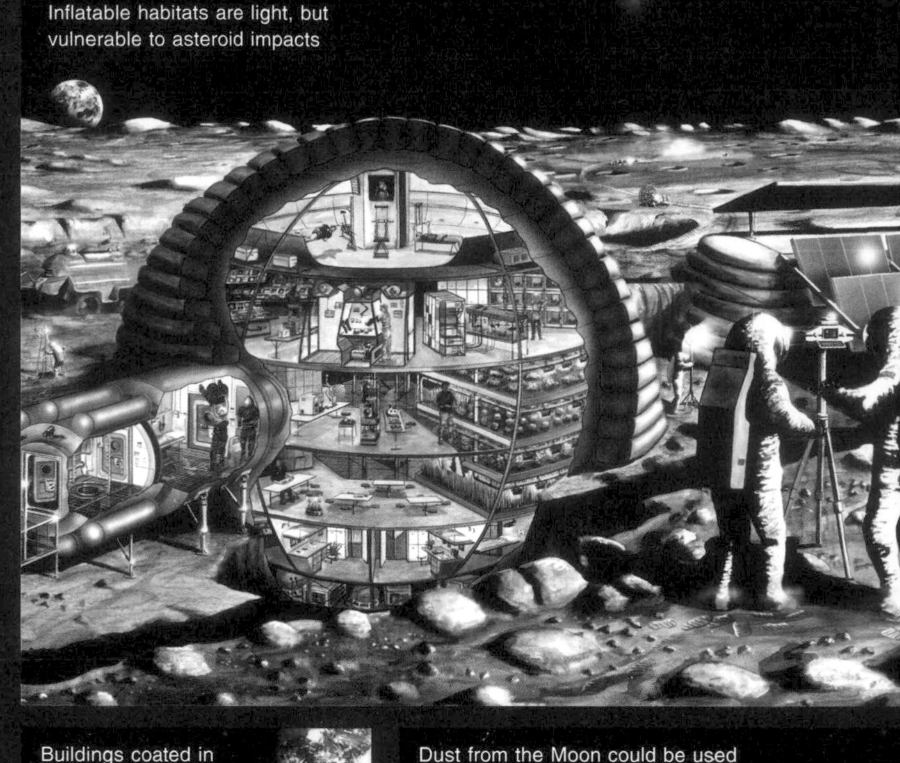

Inflatable habitats are light, but vulnerable to asteroid impacts

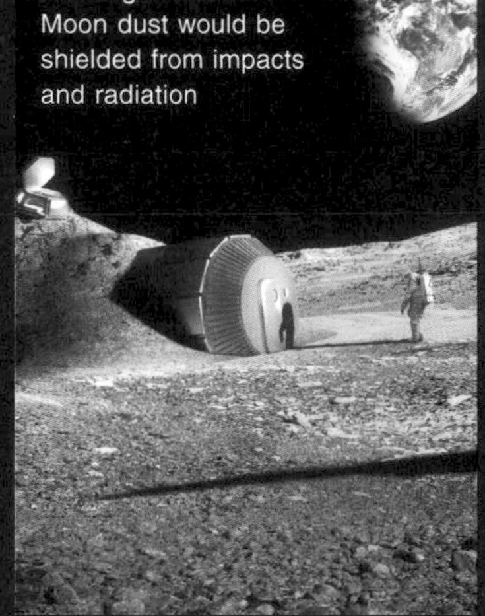

Buildings coated in Moon dust would be shielded from impacts and radiation

Dust from the Moon could be used as a material for 3D printing

Excavation equipment would need to resist the damaging effects of fine dust particles

Craters
Craters near the poles could provide protection against solar wind.

Permanent shade
The north pole is smoother than the south pole, but parts of it are in constant shadow.

Helium-3
Solar winds have left rich helium-3 deposits near the equator, providing a potential source of clean energy.

Where to build?

Choosing the right spot could mean the difference between success and failure

Smooth terrain
The surface near the equator might be easier to land on, but the temperatures here vary by hundreds of degrees.

NEAR SIDE

FAR SIDE

Lava tubes
Caverns beneath the surface of the Moon could provide shelter from radiation, space weather and temperature changes.

Sunlight
The equator is in darkness for 14 days at a time, but some places near the poles are in near constant sunlight.

Water ice
There is frozen water locked away near to the Moon's north and south poles.

Location, location, location

The Apollo missions landed close to the Moon's equator, where the surface is smooth and entering orbit is easy, but these regions have serious problems with temperature control.

The Moon turns on its axis once every 28 Earth days, so daytime at the equator lasts for two weeks, and temperatures climb to more than 100 degrees Celsius. For the other two weeks, the same spot is plunged into total darkness and the surface cools to 150 degrees below freezing. These wide fluctuations could pose real problems for buildings and equipment, and with sunlight absent for days at a time, solar power would be intermittent. Facing head-on to the Sun and with little in the way of atmosphere, the equator is also blasted by radiation and solar winds.

At the poles, night and day are less dramatic. The surface is rougher, but certain areas receive sunlight for most of the year, and the temperature remains more stable at around zero degrees Celsius. There is also water ice trapped at the poles, which could provide gases, fluids and even rocket fuel.

One promising location is Shackleton Crater, which is found at the Moon's southern pole. It receives sunlight for around 80 per cent of the year, which could provide a near constant source of electricity from solar panels. Building a base near the equator would be more challenging, but underground habitats could provide enough protection in more exposed locations.

Lava tubes like the Marius Hills pit could offer ready-made shelter from temperature fluctuations, solar wind, radiation and surface dust.

© ESA; NASA; REX

What would a lunar colony look like?

The Moon is not a safe place for humans; the base will be essential for survival

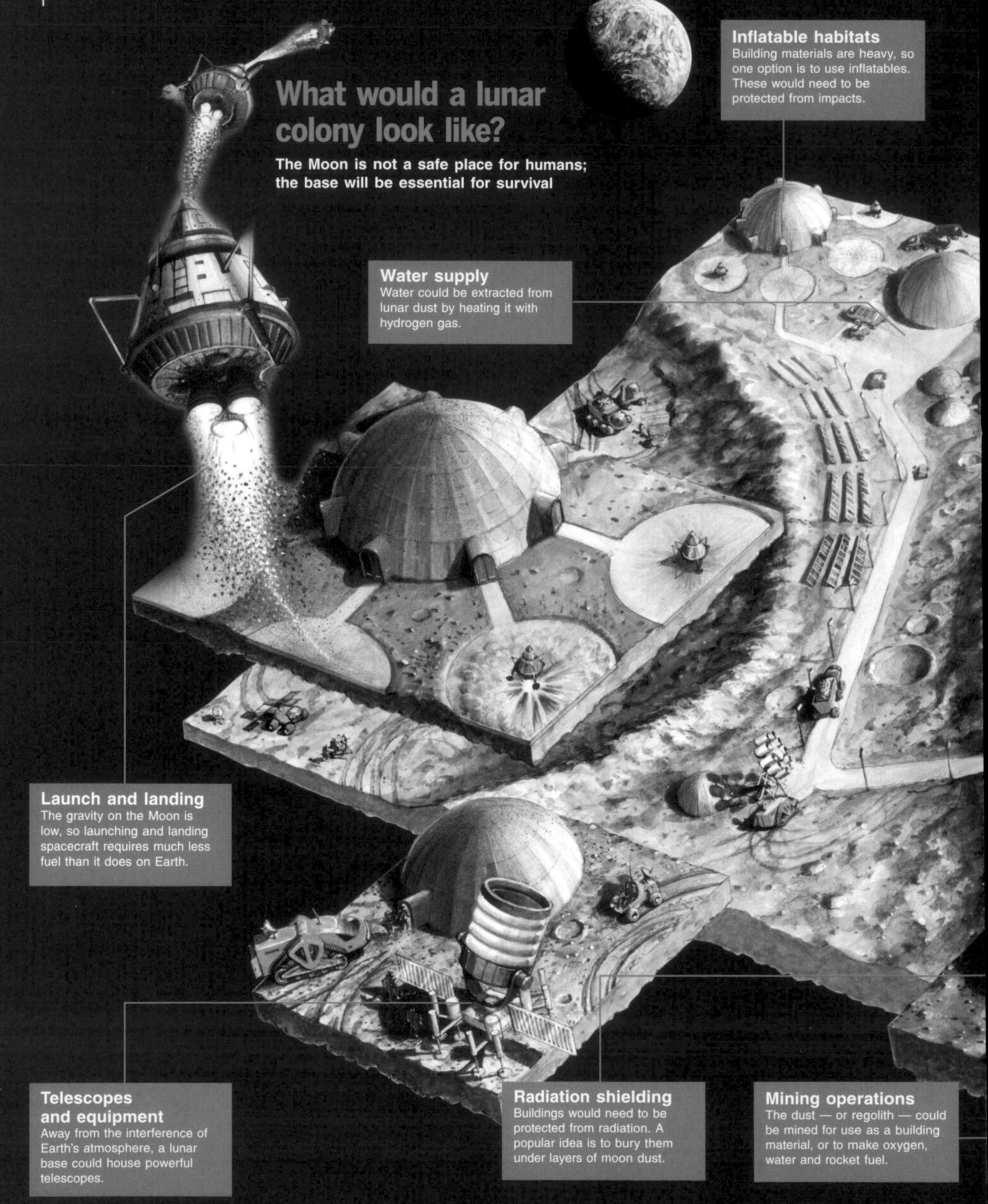

Inflatable habitats
Building materials are heavy, so one option is to use inflatables. These would need to be protected from impacts.

Water supply
Water could be extracted from lunar dust by heating it with hydrogen gas.

Launch and landing
The gravity on the Moon is low, so launching and landing spacecraft requires much less fuel than it does on Earth.

Telescopes and equipment
Away from the interference of Earth's atmosphere, a lunar base could house powerful telescopes.

Radiation shielding
Buildings would need to be protected from radiation. A popular idea is to bury them under layers of moon dust.

Mining operations
The dust — or regolith — could be mined for use as a building material, or to make oxygen, water and rocket fuel.

Oxygen
Water extracted from the lunar surface could be split into hydrogen and oxygen using a technique called electrolysis.

Glass roads
Microwaves could be used to melt the dust on the surface of the Moon to produce smooth, tough roads.

"Only a handful of people have visited the Moon's surface, and the longest stay lasted three days"

Food
Farming resources would need to be transported to the Moon, but waste could then be recycled to keep plants growing.

Flatpack buildings
Buildings could be constructed using geometric frameworks shipped in pieces from Earth.

Home away from home

Humans have been living in space since the 1970s, falling around the Earth inside orbiting space stations like Salyut, Almaz, Skylab, Mir and the International Space Station (ISS), but no one has been away from home for longer than a year, making the long-term success of space colonies hard to predict. Over 200 astronauts and cosmonauts have lived on the ISS, and by monitoring them closely we have learnt a lot about the effects of microgravity on the human body, but the Moon is a different environment. Only a handful of people have visited the surface, and the longest stay lasted for only three days.

The Moon has one-sixth of the Earth's gravity, and comes with its own unique challenges. The dust that coats the surface could prove one of the most difficult problems to overcome. During the Apollo missions, the sharp particles found their way into equipment, through vacuum seals, and even inside spacesuits, irritating the eyes and lungs of the astronauts.

Permanent settlements on the Moon will only be possible with proper protection

Preserving Our Space History

As we strive to explore and expand beyond the sphere of our own planet, we're leaving a growing trail of artefacts with vital cultural significance along the way. Should we preserve our space history – and how?

Written by Paul Cockburn

Few people — except for certain conspiracy theorists — are likely to dispute that genuine history was made on 21 July 1969 when Neil Armstrong made the first ever human footprint on the surface of the Moon. Even today, the technological achievements of the United States' space agency NASA and its Apollo space programme — fulfilling President Kennedy's 1961 commitment to "landing a man on the Moon and returning him safely to the Earth" before the end of the decade — remains a monumental landmark of human achievement.

So it surely follows that the location of those first human steps on another stellar body are of historic importance too, and should be preserved for the future. Yet it's only in the last decade that the issue has really gained serious attention from archaeologists and historians. Nor are we just talking about the Apollo missions.

The still relatively new field of space archaeology potentially covers every aspect of humanity's exploration and use of space, from the probes we've sent out to the edge of our Solar System to the geostationary satellites on which so much of our everyday lives now depend — as well as the launch systems and technology used to get them beyond the Earth's atmosphere.

"All of this material culture represents the technological stage of humankind at a particular stage of our evolution and history," insists Dr Beth O'Leary, Professor Emerita in anthropology at New Mexico State University. "The artefacts and sites are symbolic of the political, social and economic history of the world during this early space age. Although scientific in nature, space archaeologists look at the meaning and significance of objects and sites. I rank our first lunar landing right up there with the discovery of fire."

So if your idea of an archaeologist is either Harrison Ford with a fedora and a whip, or of eccentric people using small trowels and brushes to uncover artefacts and skeletons buried for thousands of years, then you may well be surprised to learn that some are now turning their attention to space.

Interview Bio

Dr Beth O'Leary

Professor Emerita in anthropology, New Mexico State University
Beth has been involved with the cultural heritage of outer space for the last 14 years. With NASA support she has investigated both the archaeological assemblage and the international heritage status of the Apollo 11 Tranquility Base site.

"A good definition of archaeology is the study of the relationships between patterns of material culture and patterns of human behaviour," Dr O'Leary says. "It sets no temporal or spatial limits. It can be done in all times and in all places. Space archaeology and heritage is the study of material culture that is relevant to space exploration that is found on Earth and in outer space."

So why the growing concerns about preservation? "Archaeology, unlike other disciplines, cannot exist without the material remains of human behaviour," she adds. "The exploration of space has a material culture that is as relevant to the development of human culture and human evolution as the earliest stone tools 2.5 million years ago.

"Study of the material record, which includes examples of technology, is an essential way to understand how and why humans create and adapt technology to explore even such extreme environments such as space. If the material remains are absent, destroyed, looted or not preserved, we lose the ability to understand humans from a unique and important perspective."

It's less than 30 years since Brown University archaeologist Richard Gould first proposed that aircraft wrecks might provide important information, laying the foundation for systematic archaeological studies of human flight. As recently as 1993, University of Hawaii anthropologist Ben Finney — who has spent much of his career examining the technology and techniques used by early Polynesian colonisers in the Pacific — suggested that we should start thinking about Russian and American sites on the Moon and Mars. While archaeologists are currently unlikely to be in a position to conduct proper fieldwork at those sites in the foreseeable future, that doesn't mean they — or unscrupulous treasure hunters — never will.

Lunar Historical Park?

On 8 July 2013, United States Congresswoman Donna F Edwards introduced a bill in the House of Representatives — numbered HR 2617 — which proposed to "establish the Apollo Lunar Landing Sites National Historical Park on the Moon", in part to "expand and enhance the protection and preservation of the Apollo lunar landing sites and provide for greater recognition and public understanding of this singular achievement in American history".

The bill also committed the US Department of the Interior to submit the sites to UNESCO (United Nations Educational, Scientific, and Cultural Organization) for designation as World Heritage Sites. However, the bill was not brought before the Congress for a vote, in part because of criticism that it could be seen as a claim of US sovereignty over the Moon.

We know that Armstrong's footprints on the Moon and those made by fellow astronaut Buzz Aldrin are still there; in 2012, NASA's Lunar Reconnaissance Orbiter (LRO) took pictures of the site from just 24 kilometres (15 miles) above the surface and confirmed it was essentially as the astronauts had left it — hardly surprising given that, unlike Earth, there's no atmosphere to erode or disturb anything. At the moment, only an extremely unlucky meteorite strike is likely to destroy those unique first human footprints in the lunar dust.

However, because we're potentially on the verge of a new space age that will see both commercial and government missions to the Moon, Dr O'Leary believes that ways must be found to evaluate and preserve such "critical phases of space exploration". We can't, in other words, always rely on the Apollo sites being safe from future human activity simply because they're 384,400 kilometres (238,855 miles) away from Earth.

Ironically enough, the biggest challenge space archaeologists face in terms of preservation is legal. "It is a question about who owns space," Dr O'Leary explains. "According to the Outer Space Treaty of 1967, no nation or state can claim the surface of the Moon or other celestial bodies, and space is regulated as a place for peaceful purposes. The Outer Space Treaty also states that the nation or state that puts objects and/or personnel in space or on other celestial bodies maintains responsibility and ownership of such. There is a whole field of space law that, by multinational and multilateral agreements regulates (albeit in a piecemeal fashion) the launching and positioning of space vehicles such as satellites. In the research I have done, these agreements and treaties do not address the preservation of cultural resources in space or on any celestial body.

"In 2000, during my work on the Lunar Legacy Project, we contacted NASA and the Keeper of the National Register of Historic Places to

Luna 2

Mare Serenitatis, the Moon

First lunar landing

The USSR's robotic Luna 2 probe was the first human-made object to 'land' on another celestial body (near the Aristides, Archimedes, and Autolycus craters), albeit at speed, and presumably leaving its mark with an impact crater. Luna 2's known instrumentation — including geiger-counter, magnetometer and micrometeorite detectors — confirmed that the Moon had no appreciable magnetic field or radiation belts. But, given this was during the Cold War, was any other technology on board?

Deep Space 1

Solar orbit

NASA's first ion-powered rocket
Deep Space 1 was the first technology demonstration probe created by NASA's New Millennium Program; in addition to its payload equipment, the technology demonstrated by the craft included its own xenon ion engine. The craft's ion engines were shut down on 18 December 2001, though its radio receiver was left on in case any future contact was required.

Helios probes

Solar orbit

Fastest, closest flyby of the Sun by any spacecraft
Helios-A and Helios-B (aka Helios 1 and Helios 2) were a joint venture between what was then West Germany and NASA to study solar processes nearer the Sun, both passing inside the orbit of Mercury. Launched in December 1974 and January 1976 respectively, the probes continued to send data up to 1985 and, while no longer functional, remain in their elliptical orbit around the Sun.

nominate the Tranquility Base site on the Moon as a National Historic Landmark," she adds. "The response from both was that making these part of the US preservation system would be perceived by the international community as a claim of sovereignty over the surface of the Moon, and the Keeper further felt that federal historic preservation law did not have jurisdiction over sites on the lunar surface as they were not on American soil."

In 2011, NASA recognised the need to address preservation on the lunar surface and included Dr O'Leary in the team, which subsequently issued NASA's Recommendations to Space-Faring Entities. However, while several commercial operations have agreed to its guidelines, they are not legally binding. "Space archaeological sites, as well as objects, fall into a grey legal area as preservation was not seen as an issue when national and international laws and agreements were drafted," explains Dr O'Leary. "Space and celestial bodies are perceived as a commons."

Another challenge is that not everything is of archaeological interest, even though most of it is located between 160 kilometres (99 miles) and 2,000 kilometres (1,243 miles) away in low Earth orbit.

From flecks of paint and metal fragments, to spent rocket stages and old satellites, this orbital space debris — a potential danger to current and future space missions — also contains 'significant' objects that warrant protection, like Vanguard 1, the US satellite launched in 1958 that's currently the oldest human object still in orbit. "We do not currently have a cohesive way of evaluating the significance of the 'space junk'," accepts Dr O'Leary, "which is an obvious first step in considering whether to discard or preserve it. Not everything can and should be preserved, or be considered 'significant' and warrant protection. That doesn't happen on Earth. Frequently in the presence of redundant debris, the debris is sampled and that sample is part of a preservation strategy. Space junk is a serious problem in low Earth orbit, but not all space junk is just garbage; there are some historic spacecraft still there that represent the technological, political and social exploration of space. The evaluation of this cultural heritage has barely been considered."

Some space missions are potentially luckier; there are fundamental mission reasons for locating the new James Webb Space Telescope in an elliptical orbit about the second Lagrange point (so that the

Viking 1 lander

Chryse Planitia, Mars

First successful lander on Mars

Viking 1 consisted of an orbiter and lander, with cameras and sensors to search for life and analyse the soil and atmosphere. Contact was lost after a faulty command was sent in 1982; the associated Orbiter had already been moved into a higher orbit (preventing a crash until at least 2019) and shut down.

Voyager 1 & 2

Outside the Solar System

Most distant human objects

According to Peter Capelotti, a professor of anthropology at Penn State University, there remains a tiny chance these interstellar probes might become "archaeological representatives of Homo sapiens to the rest of the galaxy". Yet they also remain a snapshot of humanity at the time of their launch in 1977, thanks to their gold-plated audiovisual discs.

Sputnik

Baikonur Cosmodrome, Kazakhstan

World's first artificial satellite

The actual Sputnik satellite launched by the Soviet Union on 4 October 1957 burned up on re-entry some three months later, but various replicas and back-up models do survive — ironically, many in the US, the USSR's Space Race and Cold War rival. However, the physical results of all the engineering know-how and technological developments that made the launch possible can still be found at the world's original space launch facility — which remains heavily in use today.

"I rank our first lunar landing right up there with the discovery of fire"

gravitation forces of the Sun, Earth and Moon will hold it in a stable location), but this could also ensure the safe survival of the infrared telescope well beyond its planned lifespan. However, 'future curation' to preserve historical artefacts has yet to be included in any space mission. "If there are decisions to be made about what should be destroyed or removed it should be done so that 'precious artefacts' should be left in their natural setting, given the risk factors for collision or damage," says Dr O'Leary. "These decisions need to be informed by the field of space archaeology. Governments, commercial entities and researchers all need to get involved in preservation decisions," she adds. "There needs to be multinational and multilateral cooperative decisions by those space-faring entities; the USA, Russia, Japan, ESA and China have separate and mutual interests in preserving this legacy of wonderful things in order to allow for the study of the space age."

However, if you're keen to become the first space archaeologist, you are unfortunately too late. In November 1969, Apollo 12 astronauts Charles 'Pete' Conrad and Alan Bean landed their Lunar Module in the Oceanus Procellarum (Ocean of Storms), just a few hundred feet from the crater in which the unmanned probe Surveyor 3 had soft-landed approximately two and a half years earlier. As part of their mission, the astronauts unintentionally became the first practitioners of extraterrestrial archaeology when they found the remnants of the device, carefully photographed the impressions made by its footpads and then removed the probe's television camera, remote sampling arm, and pieces of tubing — items that were then bagged, labelled and stored alongside the mission's geological samples.

In 1972, NASA published Analysis of Surveyor 3 Material and Photographs Returned by Apollo 12, which focused on the ways the retrieved components had been changed by the craft's voyage through the vacuum of space. The most surprising finding was evidence of the bacteria Streptococcus mitis on part of the camera. Obviously this was not of extraterrestrial origin and, eventually, it was concluded that someone had sneezed on the device.

So it could be said that space archaeology's first great discovery could well be that a bacteria had travelled to the Moon in an alternating freezing/boiling vacuum for two and a half years, and returned promptly to life upon reaching the safety of a Petri dish back on Earth. It would seem that not even the hostile vacuum of space could stop humans from spreading a sore throat!